ENVIRONMENT-FRIENDLY PRODUCTS
—ADAPT GREEN NOW

ENVIRONMENT-FRIENDLY PRODUCTS —ADAPT GREEN NOW

Green Marketing

DR. APARNA P. GOYAL

PARTRIDGE

Copyright © 2018 by Dr. Aparna P. Goyal.

ISBN: Softcover   978-1-5437-0167-8
           eBook   978-1-5437-0166-1

All rights reserved. No part of this book may be used or reproduced by any means, graphic, electronic, or mechanical, including photocopying, recording, taping or by any information storage retrieval system without the written permission of the author except in the case of brief quotations embodied in critical articles and reviews.

Because of the dynamic nature of the Internet, any web addresses or links contained in this book may have changed since publication and may no longer be valid. The views expressed in this work are solely those of the author and do not necessarily reflect the views of the publisher, and the publisher hereby disclaims any responsibility for them.

Print information available on the last page.

To order additional copies of this book, contact
Partridge India
000 800 10062 62
orders.india@partridgepublishing.com

www.partridgepublishing.com/india

CONTENTS

Chapter 1	Introduction	1
1.1	Background	1
1.2	Green Products Awareness and Acceptance Marketing	3
1.3	Eco- Fmcg Sector	6
1.4	Neuro Awareness and Acceptance Marketing	8
1.5	Research Outline	8
Chapter 2	Methodology	11
3.1	Objectives	11
3.2	Research Method Used	12
3.3	Survey method and administration	13
3.4	Research Ethics	18
Chapter 3	Data Analysis	20
4.1	Respondent Profile	20
4.2	Reasons For Not Purchasing	22
4.3	Store Preference for Purchasing Green Textile Products	24
4.4	Store Preference for Purchasing Green FMCG Products	27
4.5	Knowledge about Environment Friendly Products	30

	4.6	Perception about Environment Friendly Products 34
	4.7	Difference in Perception about Environment Friendly Products among Male and Female 39
	4.8	Extent of Purchase of Environment Friendly FMCG Products........................... 53
	4.9	Extent of Purchase of Environment Friendly Textile Products 56
	4.10	Factor Impacting the Green Products Purchase... 59
	4.11	Impact of various factors on Green Product Purchase Decision 71
	4.12	Difference among Conventional Awareness and acceptance marketing and Neuro-Awareness and acceptance marketing... 72
Chapter 4		Discussion 76
Chapter 5		Implications, Limitations and Recommendation for Further Research 81
References 85		
Annexure I		Questionnaire 93
Curriculum Vitae.. 103		
Annexure II		Published Papers 117

Dedicated to my husband Prashant and sons Pranav & Parth for providing me unconditional support

Chapter 1

INTRODUCTION

1.1 Background

The growing social and regulatory concerns for the environment lead an increasing number of companies to consider green issues as a major source of strategic change. In particular, this trend has major and complex implications on the awareness and acceptance marketing strategy of the products and on its product innovations. Even though it has increased eco-awareness of Indian customers during the past few decades, there are some barriers to the diffusions of more ecologically oriented consumption and production styles. Therefore, companies are increasingly recognizing the importance of green awareness and acceptance marketing concepts and finding new ways like neuro awareness and acceptance marketing to market eco-friendly products.

Green awareness and acceptance marketing incorporates a broad range of activities, including product modification, changes to the production process, packaging changes, as well as modifying advertising. (Michael Jay Polonsky,1994). Today, awareness and acceptance marketing parishioners of FMCG and Textile sector per se in India use environment friendly packaging and modify the products to minimize the environment pollution. However there is a big argument among the awareness and acceptance marketing philosophers regarding attractiveness of green product to customers in developing country like India. With this background, the researchers have made attention on attractiveness of green awareness and acceptance marketing strategies in India. The Main objective of the study is, to evaluate consumer attitudes and perception regarding eco-friendly FMCG and Textile products and to know how much the neuro-awareness and acceptance marketing could help the marketers in promoting sales and increasing revenues by using harmless to environment eco-label (CSR). The key to answering about the high level of importance being placed on sustainability is related to the very existence of human race; thereby implying that if the society, corporations, industries, government, law enforcement, consumers and everybody else still ignores to the fact that environment sustainability is a mandate to existence of ecology and the ecological system. The fact of the matter is that we are already late in adopting the green and if we do not do it now, our own existence is questionable. We cannot afford to play with nature any more, we have to give back, work back towards sustaining the ecological balance. There is still by large the notion that everything is running smoothly and it is ok to move as it is in the current situation. Whatever role we are playing in the society, we need to critically understand that saving the environment is a composite effort, where each and every member of the system plays an significant role in maintain the ecological balance. Every human being has to contribute towards the interrupted and smooth functioning of the environment. In fact, all living beings are equally important in contributing towards maintenance and sustenance of the ecological full cycle. This is the time to think that either exists together or we do not exist at all in near times to come. Saying that, it is remarkably clear that the green sustainability concept means moving towards better resources, non-exploitation of the natural wealth and pacing change in a manner it causes positive influence on ecosystem, even if it is a little change that is brought out. Green Marketing is not to be taken in isolation from the other arenas that are working towards environment sustainability. Having said that, we have to treat our surrounding environment as a fragile one, where we cannot afford to be careless, thus causing any tiniest imbalance because it can and it will create repercussions that will extend like any chemical chain reaction will which in turn affect the entire society.

1.2 Green Products Awareness and Acceptance Marketing

The concept of environment sustainability has been a concern for human race. It did not create a serious impact on the consumer and marketer till the environment degradation began affecting the mankind. According to the various philosopher's eco-green awareness and acceptance marketing is a philosophy that has been continuously involved in the protection of society.

Eco-consumption creates awareness and raises voice by different means from the complete process of manufacturing or production until the use by the consumers or end-consumers with the supreme objective of motivating the customers and consumers in Business to Business or Business to consumer markets for purchase, use and consumption of naturally produced products, be it food, beverages, textiles, home products or any other good, product, brand or service that minimize the decay of ecological balance. Green awareness and acceptance marketing is all about saving the society and minimizing the ecological balance to help boost the usage and consumption of purely organic goods, products and brands. The technology involved in going Green is all-pure that helps improve the fertility of soil, avoid use of pesticides and insecticides which are harmful for living beings, energy creation and the conservation by alternate means, which do not pollute the environment. The objective is to have a win-win situation such that the eco-system is sustained and more usage of recyclable materials and resources is promoted.

The study attempts to find out the impact of promotional activities used by the manufacturers and marketers in order to bring high eco-friendly recall and recognition of the organically produced products. Green awareness and acceptance marketing has a negative consequence on the overall well-being of human beings. Ultimately, the practices that degrade the natural environment are bringing back diseases in multifold.

Now, people are insisting pure products – edible items, fruits, and vegetables based on organic farming. The number of people seeking vegetarian food is on rise.

A workshop recently held in Mumbai about reviving organic family cultivation by Karan Manral (2017), highlighted that more people are developing interest and desire in green farming because it makes them not only self-sufficient but also takes care of constraints of finances and high costs. By this there has been high rise in consumption of naturally grown pure vegetables and fruits, instead of the ready-made canned/ tetra packs used because of time constraints.

In the beginning, the families at Mumbai and Hyderabad (India) were skeptical because they were not ready to wait for years for the soil to be completely natural, free from harmful chemicals. Upon continuous drives by the environmentalists, NGO's and other not for profit organizations, the natural farms began to rise with demand for fresh and healthy alternatives. "We have seen tremendous growth and momentum in the past five years in the [organic farming] sector," says N Balasubramanian, the chief executive of 24 Mantra Organic. "The awareness about organic products has increased manifold among the consumer and the demand for organic products has increased tremendously." Government interventions played a huge role in ecological sustainability, as the government took the pledge, "To increase crop yields in rain-fed areas, which account for nearly 55 per cent of the country's arable land, organic farming is being promoted". Few years back one school of thought prevailed – that is we take care of the main or crucial aspects of green environment through sustainable marketing, we can very well tackle the bottlenecks that are creating havoc in our ecology. Numerous strategic plans and tactics brought forward and implemented. It is from the last few years that the credibility of the marketing organizations have been questioned. So to say, most of these industry set-ups were over-emphasizing their efforts and contribution towards the greener ecology, than what was actually happening. It is when the government intervention helped and the clear-cut ecologically sustainable strategic guidelines were made clear to all the organizations. Various local bodies, NGO's and groups came forward in spreading the awareness through concerted efforts in these campaigns. Sooner there were stringent rules with checks and balances, which were communicated and implemented. This led to the encouragement and motivation among the opinion leaders causing a halo effect, thus leading to a larger emergence of population who reached out to help in attaining the objectives of greener scenario. Compliance observed right from the raw material purchase to the purchase of product by the end-customer to its divestment. The motto of using the biodegradable materials at various levels of product production soon became popular. Stakeholders, at large believed in the concept of "recycling" and tried to apply them right from transportation, electricity saving, business waste recycling, low carbon emissions, use of alternate sources of energy like solar energy, wind energy and non-exploitation of natural resources present in our ecological syatems.

1.3   Eco- Fmcg Sector

"Kerala (Southern India) is one of those States in India where land resources are put to more intensive use than anywhere else, mainly because of the low per capita availability of land in the State. Out of a total geographical area of 38.86 lakh ha, little over one fourth was under forests, and one tenth of it was put to non-agricultural use. For example, while it's probably easier to prevent insects and other fungal/bacterial diseases from getting to your crops, other beneficial ones probably get eliminated too. FMCG uses the benefits of green manuring. More advanced practitioners adopt methods of agro-forests and green plantations and apply to vegetables too. The consumers are more than ever informed about the environmental friendly products and what eco-friendly consumerism is. Corporations today are producing products, brands and style images that are ethically responsible, which acknowledged the mass consumers about fair trade. Furthermore television and print media is also developing advertisement materials, blogs and articles that contain information about ecofriendly textiles. Ethical consumption means buying goods which are produced considering the prevention of resources, environment and social issues or in other words goods exchanged with fair trade. Due to the availability of information, consumers have become more and more stimulated to purchase the eco-friendly apparel to play their part in the good cause of rising for the environmental and social concerns. This research is intended to carry out a focussed group survey based on gender, age, education and geographic location of the participants to collect data and analyse the extent of implication that eco-friendly consumerism was able to transfer in the real life decisions of the consumers. The survey questionnaire will also give us information about the factors and preferences of consumers that act as a supportive attribute or as a barrier in order to purchase or forgo the eco-friendly apparel. Life Cycle Assessment (LCA) is an important tool for achieving environment friendly products. Many companies have recently committed to apply LCA for their manufactured products. An LCA inform the consumer about the environmental effects and consequences including water use, CO_2 emission and the eutrophication process during the entire life of the product. It leads to reduce of water use, CO_2 emission as well as the level of eutrophication process if it followed strictly and in a systematic way. Introducing LCA of the garment products will promote green labelling of garment products.

1.4 Neuro Awareness and Acceptance Marketing

Finally even if previous studies were made as well in industrialized countries as developing ones, we found almost negligible studies about use of neuro-awareness and acceptance marketing techniques for awareness and acceptance marketing Green FMCG and Textile Products.

1.5 Research Outline

This thesis is divided into five chapters. The first chapter is reserved for the introduction of study background, Green Awareness and acceptance marketing, Neuro Awareness and acceptance marketing and FMCG & Textile Sector. This is just to give a snapshot of the subject matter and premise of the study. Chapter 2 presents an in-depth discussion and literature review on the Green Products Awareness and acceptance marketing and Neuro-Awareness and acceptance marketing. Chapter 3 is designed to present

Research Methodology: This is to illustrate the research questions and the hypotheses. In addition, the chapter illustrates how the data will be collected, what sampling techniques and statistical methods will be used. Chapter 4 evaluates the empirical results by analyzing the findings of the individual research questions and hypothesis taking into consideration the various objectives. Subsequently, Chapter 5 presents the finding by discussing the research questions and implications of the results along with conclusion. Lastly, Chapter 6 presents the limitations of the study along with the directions for the future research.

The cursory overview of the chapter refers to the evolution of environment friendly awareness and acceptance marketing in various sectors, in general and FMCG and Textiles in particular. The chapter emphasizes on the role of consumer, society and Govt. in NCR. At the end of the chapter it may be asserted here that a global economy, in which world class quality is the ticket to success, increased awareness in the society, calls for more ethical conduct and to keep things interesting. As trustees of society's precious human, material, financial, and informational, environmental resources, marketers as well as consumers, hold the key to a better clean environment in the world. If one goes for a retrospective journey of the last century, the tide has been running strongly in the favor of the virtues of environmental stability and protection that have been found unquestioned and seemingly self - evident for decades. The lifetime contributions of many authors, researchers and society were robust but they have gone largely unchallenged and unexamined. The perception of customers in India must be strengthened by emphasizing on deteriorating state of earth and involve society to strongly believe that the battle for solving problem of tomorrow's state of ecological environment, if not be fought today with strategies and weaponry of awareness and acceptance marketing management and related domains. The environmental scenario has been turned upside down now, most of the people in manufacturing or awareness and acceptance marketing of green products are now experiencing the increased customer awareness and expectation towards being green and helping in saving ecological operating balance with considerable societal awareness and acceptance marketing and related tactics that were designed to cope up for altogether different reality. Many, if not most, of what today's Green marketers formerly grew up believing to be unquestioned assets as their products / services which have now have become liabilities because of lack of pro-active approach in adapting to safer, bio-degradable and eco-friendly products. The growing purchasing power and rising influence of the social media have enabled Indian consumers to splurge on good things. The Indian consumer sector has grown at an annual rate of 5.7 per cent between FY2005 to FY 2015. Annual growth in the Indian consumption market is estimated to be 6.7 per cent during FY2015-20 and 7.1 per cent during FY2021-25.

India's e-commerce market is likely to reach US$ 38 billion (Rs 252,700 crore) in 2016*. The online retail sector in India is expected to be a US$ 1 trillion (Rs 660,000 crore) market by 2020@. Amazon expects India to become its quickest market to reach US$ 10 billion in gross merchandise value (GMV) and to become its largest overseas market surpassing Japan, Germany and the UK.

Experience may have been the dominant currency of yesteryears, now that experience along with knowledge of growing and changing customer perceptions about their ethical duty and moral behaviour, is the new name in digital world as well, of the currency, where people are connected mouse to mouse. In times to come flexibility not compromise, agility not stability, skepticism not certainty and an insatiable appetite for adapting to contemporary customer need and want, will make knowledge and learning the recurring theme in Indian Markets.

An attempt has been made in the next chapter to trace the genesis of emergence of eco-friendly products in India.

Chapter 2

METHODOLOGY

3.1 Objectives

- To understand the consumer's perception towards Eco-friendly/green products.
- To study the extent of purchase of Eco-friendly/green products.
- To explore the factors affecting the purchase of Eco-friendly/green products.
- To find the extent of impact of identified factors on the purchase of Eco-friendly/green products
- To identify possible differences in terms of inviting attention, creating Interest, infusing desire and promoting purchase action among the consumers of conventional awareness and acceptance marketing and neuro-awareness and acceptance marketing method.

3.2 Research Method Used

There are five basic types of statistical analyses that often used in analyzing the research data: viz.,

- Descriptive analysis
- Inferential analysis
- Predictive analysis
- Associative analysis
- Multivariate analysis

Descriptive statistics are brief descriptive coefficients that summarize a given data set, which can be either a representation of the entire population or a sample of it. Descriptive statistics are broken down into measures of central tendency and measures of variability, or spread.

Indeed, the Neuro-Awareness and acceptance marketing is a relatively new phenomenon in India. Although the number of green user is proliferating, there is little empirical evidence to help marketers fully understand what affects consumer attitude towards Eco-friendly/green products and whether Neuro-Awareness and acceptance marketing promotes the purchase of Eco-friendly/green products. The principal sources of data for this exploratory research were:

A review of literature on Eco-friendly/green products and Neuro-Awareness and acceptance marketing in FMCG and Textile Sector was undertaken in order to define the comprehensive scope of the study and ensuring its objectivity.

Having obtained some primary knowledge of the subject matter by an exploratory study, descriptive research was conducted next. Contrary to an exploratory research, a descriptive study is more rigid, preplanned and structured, and is typically based on a large sample (Churchill & Iacobucci 2004; Hair et al. 2003; Malhotra 1999).

The purpose of descriptive research is to describe specific characteristics of consumers and to determine the frequency of purchase of Eco-friendly/green products, also to gaze their knowledge about the Eco-friendly/green products. In addition to know which brands they prefer and which products they purchase.

As many researchers have noted, descriptive research designs are for the most part quantitative in nature (Burns & Bush 2002; Churchill & Iacobucci 2004; Hair et al. 2003; Parasuraman1991).

For the purpose of this study, a cross sectional study was the appropriate technique as opposed to a longitudinal study due to time constraints, and furthermore, this study does not attempt to examine trends.

3.3 Survey method and administration

To develop questionnaire the following points are kept in the researchers mind:

- Measurement scale
- Content & wording
- Response format
- Sequence
- Physical layout

- Prepare final
- Draft
- Revise

Questionnaire distribution and administration

Reliability and validity of questionnaire

- Pre testing of the questionnaire
- Pilot test (62 users)

Method for this research was a personally administered one. This method was chosen for the following reasons (Kassim 2001):

Consumers could be contacted easily around the malls and multi-brand retail outlets and university premises.

Frequency Distribution: The researcher needs to answer question about a single variable:

- How many respondents may be characterised as Green Consumers and Non-Green Consumers?
- What is the reason behind not purchasing the Green products?
- Which particular store consumers prefer for purchasing FMCG and Textile products?
- What do consumers think about Green products?
- How often consumers purchase various FMCG and Textile products?
- What are the top five popular green FMCG and Textile products?

In this study we have selected the significance level firstly at 0.05 and then at 0.01. The former one shows that the difference is significant and the latter one shows that it is very significant.

The t distribution is similar to the normal distribution in appearance. Both distributions are bell shaped and symmetric. However, as compared to the normal distribution, the t distribution has more area in the tails and less in the center.

Factor Analysis:

It is primarily used for data reduction or structure detection. The purpose of data reduction is to remove redundant (highly correlated) variables from the data file. This study requires the factor analysis to reduce and classify the factors affecting the purchase of green products.

Regression Analysis:

The current study requires to describe the extent of impact of various factors on purchase of green products. To achive the objective, the data was further analyzed with the help of Multiple Regression Analysis. The process of estimating the value of dependent variable to a set of independent variable(s) as specified in a research problem is termed as predictive analysis. Regression analysis falls into this category and we have used it- both bi variate and multiple regression.

Regression analysis is a statistical technique that is used to relate two or more variables. Here, a variable of interest, the dependent or response variable (Y) is related to one or more independent or predictor variables (Xs). The objective in regression analysis is to build a regression model or a prediction equation relating the dependent variable to one or more independent variables. The model can then be used to describe, predict, and control the variable of interest on the basis of the independent variables.

Multiple regression analysis is an expansion of bi-variate regression analysis in that more than one independent variable is used in the regression equation. The addition of independent variables complicates the conceptualization by adding more dimensions or axes to the regression situation. But it makes the regression model more realistic because prediction normally depends on multiple factors, not just one. Everything about multiple regression is essentially equivalent to bi-variate regression except we are working with more than one independent variable. The terminology is slightly different in places, and some statistics are modified to take into account the multiple aspects, but for the most part, concepts in multiple regression are analogous to those in the simple bi-variate case.

The regression equation in multiple regression has the following form:

$$y = a + b_1 x_1 + b_2 x_2 + b_3 x_3 + \ldots + b_m x_m$$

where

y = the dependent, or predicted, variable
x_i = independent variable I
a = the intercept
b_i = the slope for independent variable I
m = the number of independent variables in the equation.

As we can see, the addition of other independent variables has done nothing more than to add bixis to the equation. We still have retained the basic y = a + bx straight-line formula, except now we have multiple x variables, and each one is added to the equation, changing y by its individual slope. The inclusion of each independent variable in this manner preserves the straight-line assumptions of multiple regression analysis. This is sometimes known as additivity because each new variable is added on. Just as in bi-variate regression analysis in which we use the correlation between y and x, it is possible to inspect the strength of the linear relationship between the independent variables and the dependent variable with multiple regression. Multiple R, also called the coefficient of determination, is a handy measure of the strength of the overall linear relationship exists among the variables. Multiple R ranges from 0 to + 1.0 and represents the amount of the dependent variable "explained," or accounted for, by the combined independent variables. High multiple R values indicate that the regression plane applies well to the scatter of points, whereas low values signal that the straight-line model does not apply well. At the same time, a multiple regression result is an estimate of the population multiple regression equation, and, just as was the case with other estimated population parameters, it is necessary to rest for statistical significance.

It is a statistical technique, which allows a researcher to study the differences among two or more groups of respondents with several variables simultaneously. In the present approach paper, an attempt has been made to know whether the chosen variables have the capacity to differentiate between and among the criterion group of respondents. This procedure also helps in determining the variables, which contribute maximally for differences among the criterion groups.

3.4 Research Ethics

All participants were made aware of research intentions and design by an introduction. Findings have been treated with the utmost confidentiality. No source, whether individual or organization was or will be correlated with specific findings or comment attributed without the consent of the originator or organization. All discussions will remain confidential in relation to other organizational participants and during the reporting of findings. All participants have the right of anonymity. It was made clear to participants that they had the right to decline to respond to any question. Participants were treated with the utmost respect and their right not to answer a question or withdraw from participation was appreciated. In order not to intrude on privacy not any further efforts were made to contact a potential participant after two attempts. Other than the offer of a copy of the completed dissertation for agreeing to participate no other participation inducements were offered. During management interviews the researcher remained cognitive of the allotted interview time and made no attempt to purposively extend the interview. As the participants interviewed for focused group interview were all high level managers with very demanding schedules no attempts were made to infringe on their goodwill with supplemental questions.

Primary data was collected with an awareness to remain objective and no data was invented. Regarding secondary data sources, all secondary data sources are given full credit for their contribution to this study. The collected data is represented honestly and the analysis is to the best of the researcher's experience and ability.

> *Based on the objectives and on chosen research design, the study was conducted on 404 respondents to study the chosen variables of the study and its implication for the Eco-friendly FMCG and Textile products in general and branded green products particular. Based on the responses and with the help of SPSS 19.0, the data were analysed and is presented in the following chapter. "Results, Analysis and its Interpretation"*

Chapter 3

DATA ANALYSIS

4.1 Respondent Profile

The research objectives were attained through the usage of relevant and appropriate tests for analyzing the data collected. The characteristics of respondents' demography like gender, age income, education, occupation, etc. have been studied because this research needed to assess from the sample population as to who has to be surveyed and how to analyze data in a meaningful form. Table 4.1 depicts the profile of the respondents taken in this research, on the various relevant parameters viz. Gender, Age, Personal income and occupation of the respondents. Out of a total sample of 404 respondents, including males and females, almost equal percentage of male (48.5%) and female (51.5%) were ensured for taking responses for the research, so that each gender gets the equal chance of participation. The similar selection criteria had been taken into account for the other variable viz. Age, in order to ensure that each person from respective equal chance of participating to meet the research standards laid down. It can also be seen from the above table 4.1 that socio demographic data like personal income and occupation as it determines if we are gathering information from the seeking group if respondents only. It makes certain the closeness of sample taken to, replicate the population. The less than Rs. 5 lacs was found to be 21.53% and 31.45%, 26.73%, 20.29% for the personal income ranges Rs. 5-10 lacs, 10-15 lacs, and above 15 lacs respectively.

It is imperative to note from the table that the, occupation of respondents had been 'Business" for 33.16%,' Services 'for 24.50% and homemaker & students as 21.28% & 21.03% respectively.

Table 4.1 Respondent Profile					
		Frequency	Percent	Valid Percent	Cumulative Percent
Gender					
Valid	Male	196	48.5	48.5	48.5
	Female	208	51.5	51.5	100.0
Age					
Valid	18-25	102	25.24	25.24	25.24
	25-35	87	21.53	21.53	46.77
	35-45	117	28.96	28.96	75.73
	>45	98	24.27	24.27	100.00
Personal income					
Valid	Dependent				
	< 5 lacs	87	21.53	21.53	21.53
	5-10 lacs	127	31.45	31.45	52.98
	10-15 lacs	108	26.73	26.73	79.71
	>15Lacs	82	20.29	20.29	100.00

Occupation					
Valid	Student	85	21.03	21.03	
	Homemaker	86	21.28	21.28	
	Service	99	24.50	24.50	
	Business	134	33.16	33.16	
Total		404	404		

4.2 Reasons For Not Purchasing

Out of 404 respondents, when asked about main reasons for not purchasing the eco-friendly products, the maximum responses (302) went towards the reason that 'They are expensive' followed by 267 respondents sighting the reason as 'Less shelf life'. Another important funding for reasons for non-purchase is lack of standardization in environment-friendly products. Other important reasons being 'Less variedly' 'Claim not trustworthy' & 'Inferior quality as compared to the chemical counterpart' with 243, 208 & 170 respondents answering respectively. Thus, it can be deduced that price, had been the most significant reason for non-purchase, while lack of standardization of green products is another reason because of which green products are over powered by their non-green products are more in demand as compared to their non-green competitive products.

Table 4.2 Reason for Not Purchasing			
Reasons	Yes	No	Total
They are expensive	202	102	404
Claim is not trust worthy	208	196	404
Inferior quality as compared to chemical counter part	170	234	404

Less variety	243	161	404
Non standardisation of products	108	296	404
Less shelf Life	267	137	404

Fig 4.2 Reason for Not Purchasing

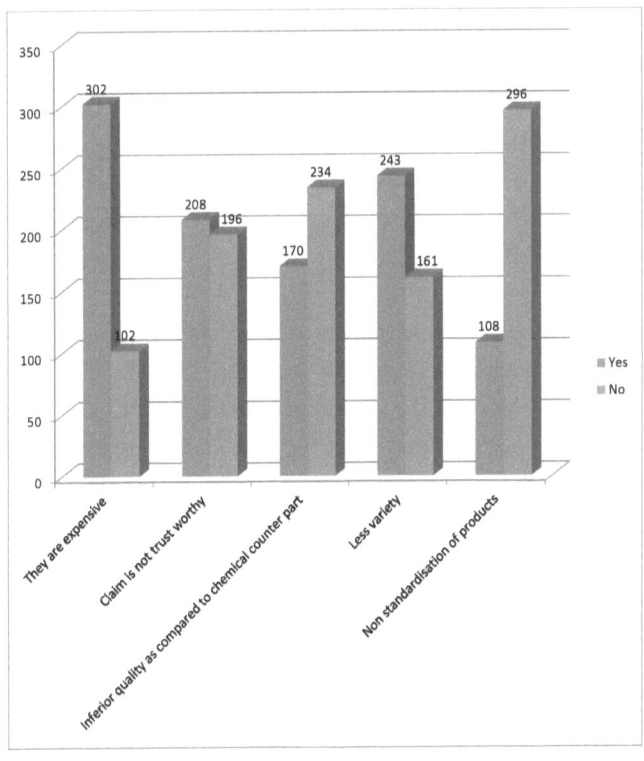

4.3 Store Preference for Purchasing Green Textile Products

A store basically comprises of a group of factors or variables that makes it distinct from the others in competition, by being a physical and emotional embodiment of a business model in practice to strategize. When respondents(404) were asked in detail about various stores or out lets they prefer when they buy the green eco-friendly her sally produced Textile products, a majority of 305 respondents answered 'Khadi Gram Udyog' as their first preference followed by 'Fab India' as an interesting preference for organic products.

It was strange to know that the respondents least preference to buy green textile products was Fibre 2 Fashion. Other stores that topped the preference charts for 'ECO' were cottage emporium 'and GOOD Earth'. Among the other preference store were 'Grassroots' Siyaram, State emporiums, earth and similar in Delhi-NCR the stores discussed in the research are generally hypermarkets & supermarkets the research has focused on. The impact of consumer perception (respondents) in buying environment friendly textile products based on the image they carry about the shopping stores. It can be concluded that a single attribute seals focus on the product being eco-friendly and that attributed is government certification revealing trust, and loyalty and so they carry that security certified label with more trust worthiness.

Table 4.3 Store preference –Textile			
Store	Yes	No	Total
Which particular store you like to purchase Environmental friendly Textile Products from?- **Khadi Gram Udyog**	305	99	404
Which particular store you like to purchase Environmental friendly Textile Products from?- **Fab India**	176	228	404
Which particular store you like to purchase Environmental friendly Textile Products from?- **Earth**	23	381	404
Which particular store you like to purchase Environmental friendly Textile Products from?- **Anokhi**	45	359	404
Which particular store you like to purchase Environmental friendly Textile Products from?- **Cottage Emporium**	98	306	404
Which particular store you like to purchase Environmental friendly Textile Products from?- **State Emporiums**	13	391	404
Which particular store you like to purchase Environmental friendly Textile Products from?- **Siyaram's**	154	250	404
Which particular store you like to purchase Environmental friendly Textile Products from?- **Fibre2Fashion**	10	394	404
Which particular store you like to purchase Environmental friendly Textile Products from?- **Good Earth**	65	339	404
Which particular store you like to purchase Environmental friendly Textile Products from?- **Grassroots by Anita Dongre**	13	391	404
Which particular store you like to purchase Environmental friendly Textile Products from?- **Any other**	23	381	404
Valid N (listwise)	404		

Fig. 4.3 Store preference –Textile

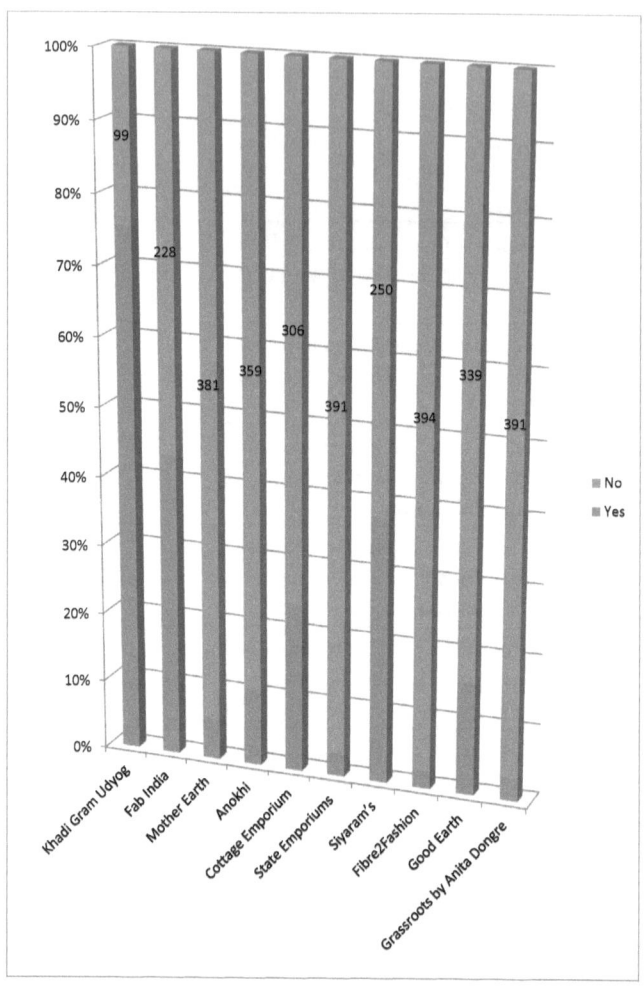

4.4 Store Preference for Purchasing Green FMCG Products

In order to understand what are the main prefers of the respondents towards purchasing the FMCG products that are ecologically safer and greener compared to their chemical or non-herbal alternatives, the respondents of size 404 were asked to marks on a dichotomous Yes-No scale, regarding their store preferences.

Consumers perceive 'green' through different varied attributes. Some of them are the raw materials used, energy consumed and the industrial waste in the form of pollution being generated. It had been found that in FMCG sector, consumers are more concerned about the genuineness of claims and promises made for their product to be environment friendly. Hence, the concept of eco-labeling came only to promote authenticated Eco-Green Products.

From the table 4.4 it can be evidently observed and concluded that out of 404 respondents, a whooping 303 gave their first choice of purchasing FMCG goods as the

Khadi Gramodyog Bhavan, because they don't perceive it as first any retail outlet for awareness and acceptance marketing and selling village/ rural industry product or Khadi Products, but the consumers perceive Khadi Gramodyog Bhavan as a non-exploitative consumer friendly and eco-supportive consumer friendly and eco-supportive organization of government of India and its products coming from rural masses are purest and safest.

The second preferences of the respondents 209 out of 404 had been undeniably the 'Patangali' Brand of Baba Ramdeveji, primarily because he is a Yoga Guru Icon and runs 'NGO' named Patanjali selling only Ayurvedic, natural and herbal products.

Baba Ramdev is also perceived to be 'Swadeshi' and make in India strong supporter. As far as its products are concerned, it was attributed that these goods have wonderful properties even when the packaging packing/design or looks of the products are innocuous.

Followed by 'Patanjali' the highest preferences had been accorded to 'Fab India' even in FMCG category, followed by 'The Body Shop' and the cottage emporiums, with and overwhelming response of 89 respondents preferring a big yes to these stores it was an insight for the researchers to note that state Emporiums, Nature's Himalaya, Eco-India and Good Earth were among the nest top 10 preferences to buy the herbal environment friendly FMCG products in various Hair-care, Skin-care, personal-care categories.

There had been a perception among the respondents that these top stores offer Quality, standardization, Government seal of authentication, status-quo because of the "Holograms or "Brand Logo" along with technical R&D collaborations with the finest research institutions of India whose aim is to enhance predictability.

It can be concluded that these preferred eco-purchase stores in FMCG are perceived to be of high Brand trust because of their mission to save the environment eco-system and visualize a chemical- free society at large.

Table 4.4 Store preference –FMCG			
Store	Yes	No	Total
Which particular store you like to purchase Environmental friendly FMCG Products from?- **Khadi Gram Udyog**	303	101	404
Which particular store you like to purchase Environmental friendly FMCG Products from?- **Fab India**	109	295	404
Which particular store you like to purchase Environmental friendly FMCG Products from?- **Body Shop**	98	306	404
Which particular store you like to purchase Environmental friendly FMCG Products from?- **Patanjali**	209	195	404
Which particular store you like to purchase Environmental friendly FMCG Products from?- **Cottage Emporiums**	56	348	404
Which particular store you like to purchase Environmental friendly FMCG Products from?- **State Emporium**	43	361	404
Which particular store you like to purchase Environmental friendly FMCG Products from?- **Organic India**	89	315	404
Which particular store you like to purchase Environmental friendly FMCG Products from?- **Natures**	87	317	404
Which particular store you like to purchase Environmental friendly FMCG Products from?- **Himalaya**	56	348	404
Which particular store you like to purchase Environmental friendly FMCG Products from?- **Eco India**	12	392	404
Which particular store you like to purchase Environmental friendly FMCG Products from?- **Good Earth**	44	360	404

Fig. 4.4: Store preference–FMCG

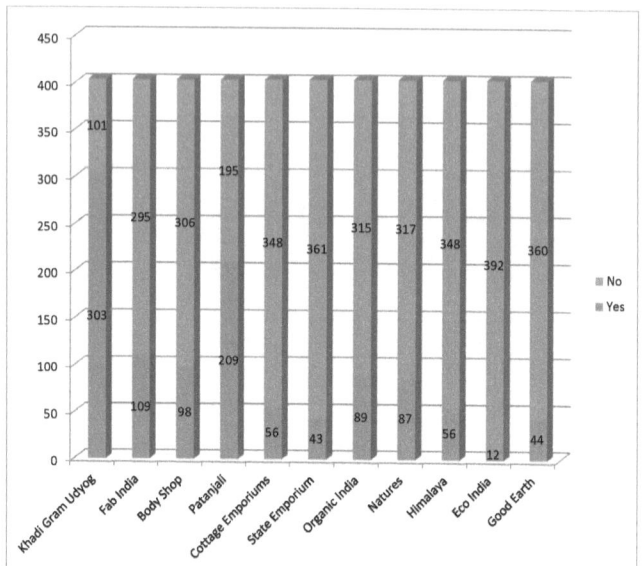

4.5 Knowledge about Environment Friendly Products

The table 4.5 depicts the extent of knowledge and awareness about 'Green' 'Herbal' 'Eco' environmentally sustainable products in the textile and FMCG sectors. In order to measure the level of awareness, interest and knowledge, the researches recorded the responses of 404 respondents on ten major aspects (In the form on statements) and calculated the mean score of each of the ten statement level of the respondents.

In a nutshell, the shopper's concurs for the pro-environmental issues, current awareness of environmental issues and it's sustainability and awareness about the availability, processes of production, ingredients, level of hazardousness, extent of recyclability reusability and bio-degradability. Other variables were the levels of toxicity, carbon & land, testing on animals, packing & packaging and disposal related concerns were recorded for analysis.

The result from table 4.5, descriptive statistics indicate that majority from the 404 respondents, were in agreement with the various questions answered to test their knowledge and understanding about what they perceive to be having in a product which is labeled 'Green' or eco-friendly or causing minimum or no hazard to the environment and to nature.

The respondents astoundingly were very understanding, aware and had knowledge that the Green product are the ones which are herbal, natural and do not disrupt nature. A man value of 4.655 very well ascertains that further, eco-pro-products are natural resources for longer sustainability the environmental resources.

Table 4.5: Eco-Product Knowledge Descriptive Statistics					
	N	Minimum	Maximum	Mean	Standard Deviation
Environmental friendly Products are the ones which- Produced by natural resources	404	1	5	4.654	.4881
Environmental friendly Products are the ones which- Help in conserving natural resources	404	1	5	4.234	.3625
Environmental friendly Products are the ones which- Complete production chain is controlled from an environmental perspective	404	1	5	2.376	.4780

Environmental friendly Products are the ones which- Are manufactured using green technology that caused no environmental hazards	404	1	5	3.865	.5009
Environmental friendly Products are the ones which- Are recyclable, reusable and bio-degradable	404	1	5	2.926	.4964
Environmental friendly Products are the ones which- Have herbal and natural ingredients	404	1	5	4.655	.3960
Environmental friendly Products are the ones which- Are non-toxic, minimize carbon emission and contain recyclable ingredients	404	1	5	3.465	.5024
Environmental friendly Products are the ones which- Do not harm or pollute the environment	404	1	5	2.098	.4964
Environmental friendly Products are the ones which- Are not tested on animals	404	1	5	1.332	.3960
Environmental friendly Products are the ones which- Have eco-friendly packaging	404	1	5	1.654	.4233
Valid N (listwise)	404				

Fig. 4.5: Eco-Product Knowledge Descriptive Statistics

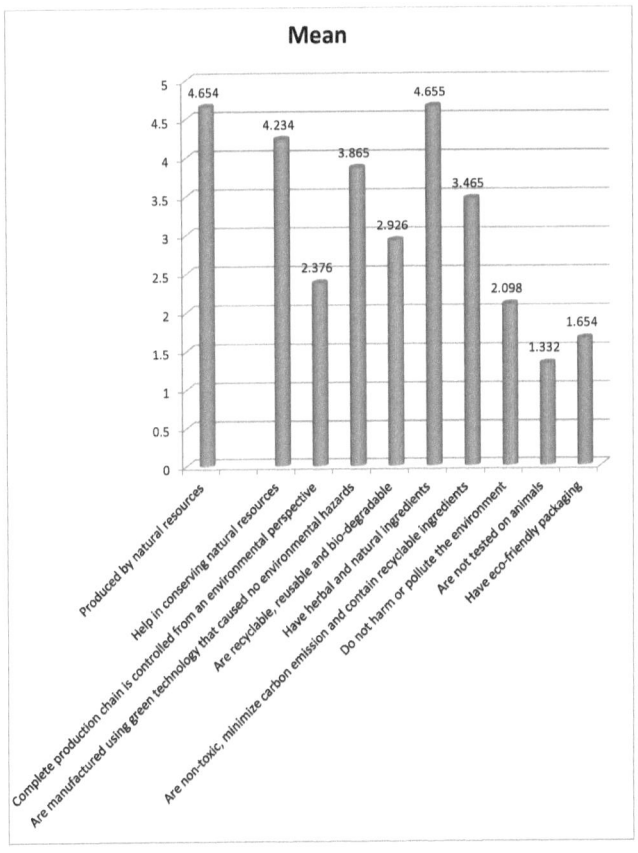

4.6 Perception about Environment Friendly Products

Table 4.6.1: Descriptive Statistics					
	N	Minimum	Maximum	Mean	Standard Deviation
The seal of quality organizations will further support & increase purchase of Eco- for greener environment.	404	2	5	4.4615	0.84673
I think that environment friendly products may cause quite less harm our environment as compared to its chemically laded counterpart	404	2	5	4.4231	0.84408
There is respect for our nature in doing so.	404	3	5	4.1154	0.93796
My ethics, Values, norms and culture motivates me to stretch a little to be ethical toward the environment I live in.	404	3	5	4.0769	0.67827
Does not harm health	404	1	5	4	0.92432
Perceived as good value for money preposition even if is priced little higher	404	1	5	4	1.14908
Employees/Customer feel that routine can be affected by implementing green concept.	404	2	5	4	0.92432
In unrelieved to buy environmentally safe products for my personal happiness and satisfaction.	404	1	5	3.9615	0.85812
Price as major factor in slow growth of sales of green products.	404	1	5	3.9615	1.19808

Statement	N	Min	Max	Mean	Std. Dev.
I think that it is ethical on my part if I support the green products because they have (i) well defined origin (ii) are natural & safe foe consumption and in disposal.	404	2	5	3.9231	0.87791
It is difficult for all companies to adapt to green initiatives compliantly	404	1	5	3.8846	0.97848
Transparency n contents of the product labeled green.	404	1	5	3.8846	1.19308
Transparency of organization using green awareness and acceptance marketing concept since long.	404	2	5	3.8462	0.77296
I think that the green product does not carry harmful ingredients like wisectiade/ pesticide/ plastic/non-biodegradable or non-recyclable input materials	404	1	5	3.8077	1.21541
I wish to purchase environment friendly nature product that do not harm the environment and society.	404	2	5	3.8077	0.83697
I am worried about the state of an environment and what holds in the future	404	1	5	3.8077	1.08007
I think green process of production respects the environmental health	404	1	5	3.7692	1.34548
I believe in the individual empowerment necessary for it.	404	2	5	3.7308	0.86159

I support the decisions taken during purchase by any peer-group/Family	404	1	5	3.7308	0.94746
Humans must live in harmony with the eco-systems/nature in order to survive.	404	1	5	3.6923	0.99588
I can purchase the green product if the packaging & labeling information provided is sufficing my clearly doubts.	404	1	5	3.6154	1.11745
I can represent my socio-economic status before my poual groups by purchasing green	404	1	5	3.5769	1.18807
I believe that with advent of latest technologies, use of eco-friendly products will increase and improve the efficiency in reduce-recycle reuse chain to support greener environment.	404	1	5	3.5385	1.37186
I take conscious decision of buying the greener attentive product for the sake of my duty towards the environment.	404	2	5	3.5385	0.6374
I make special effort to reduce the use of paper non-xcylable product to pass on this message to the society.	404	1	5	3.5385	1.08765
It improves my quality of wellbeing in life.	404	2	5	3.5	0.89225
It symbolizes the culture and values I possess	404	2	5	3.5	1.31459
the purchase of green product is for the reason that I want to do may bit for our environment.	404	2	5	3.4231	1.15492

The problem of depleting natural resources can be resolved by my message to society for green.	404	2	5	3.3846	0.92755
Trustworthiness of marketer's proofs of being eco-friendly	404	2	5	3.3846	1.08214
Advantage outweigh the disadvantage so can pay more for sake of owes and society	404	1	5	3.3462	0.9631
I am inspired by the other who works toward having a automatable society	404	1	5	3.3462	1.00261
I think my initiative can lessen the severely upset environmental balance	404	2	5	3.3462	1.0773
I am in favor of the green product that any stages have used the sustainability process from renewable success.	404	1	5	3.3077	1.14126
Natural ingredients as claimed	404	1	5	3.2692	1.35267
Keeps promise of being safe for consumption	404	1	5	3.1923	1.27772
I don't mind spending on eco/green product if it convinces me on consistency and trustworthiness of brand being herbal	404	1	5	3.1923	1.08007
I respect the wish of my other us my social group.	404	1	5	3.1538	1.06825
Quality of green product is Good.	404	1	5	3.0385	1.09654
I feel pride in explaining eh pros of eco-product, its information and how I am doing my contribution to society at larger	404	1	5	3.0385	1.23007

I believe that my simple/Small action can cause big impact towards the eco-system balance.	404	1	5	3	1.27713
I am interested purchasing eco whose by product are used for renewable initiative	404	1	5	2.9615	1.26125
I help others by persuading them to be environmentally cautious citizens.	404	2	5	2.9231	1.0766
It gives a trek message to my peer group	404	1	5	2.8846	1.0174
I am competent to pay more	404	1	5	2.7308	1.02617
I can catalyze sustainable environment by paying and enchase to the engagement & involvement & Encouragement in working as a catalyst to enable green product advantage ones non-green and hence wi	404	1	5	2.7308	1.2325
It gives me more acceptances in my primary/Secondary / treasury groups	404	1	5	2.6538	1.525
It shone my deep rootedness to cultural/historical roots.	404	1	5	2.5385	1.39988
It is my ethical obligation to pay more for environment	404	1	5	2.5385	1.01372
My culture/ Value makes me pay more for environment	404	1	5	2.3077	1.41474
I am status conscious	404	1	5	2.1538	1.3276
Valid N (listwise)	404				

According to the results about the respondent's perception on green products one would conclude that respondents feel that the seal of quality organizations further support & increase purchase of Eco- for greener environment. They think that environment friendly products may cause quite less harm our environment as compared to its chemically laded counterpart. It is because they have respect for our nature also their ethics, values, norms and culture motivates them to stretch a little to be ethical toward the environment they live in. However they strongly feel that a green product does not harm the health.

4.7 Difference in Perception about Environment Friendly Products among Male and Female

Table 4.7.1: Group Statistics					
	Gender	N	Mean	Standard Deviation	Standard Error Mean
I think green process of production respects the environmental health	male	196	3.7931	1.23911	.16270
	female	208	3.7391	1.48226	.21855
I think that the green product does not carry harmful ingredients like wisectiade/ pesticide/ plastic/non-biodegradable or non-recyclable input materials	male	196	3.9655	1.15418	.15155
	female	208	3.6087	1.27329	.18774
I think that environment friendly products may cause quite less harm our environment as compared to its chemically laded counterpart	male	196	4.3103	.92161	.12101
	female	208	4.5652	.71963	.10610
I believe that with advent of latest technologies, use of eco-friendly products will increase and improve the efficiency in reduce-recycle reuse chain to support greener environment.	male	196	3.5862	1.29824	.17047
	female	208	3.4783	1.47180	.21700

The seal of quality organizations will further support & increase purchase of Eco- for greener environment.	male	196	4.4655	.90254	.11851
	female	208	4.4565	.78050	.11508
I think that it is ethical on my part if I support the green products because they have (i) well defined origin (ii) are natural & safe foe consumption and in disposal.	male	196	3.8448	.85433	.11218
	female	208	4.0217	.90650	.13366
There is respect for our mother nature in doing so.	male	196	3.9310	.95260	.12508
	female	208	4.3478	.87477	.12898
My ethics, Values, norms and culture motivates me to stretch a little to be ethical toward the environment I live in.	male	196	3.9828	.66204	.08693
	female	208	4.1957	.68701	.10129
I am interested purchasing eco whose by product are used for renewable initiative	male	196	2.9310	1.24057	.16290
	female	208	3.0000	1.29957	.19161
The purchase of green product is for the reason that I want to do may bit for our environment.	male	196	3.1724	1.12605	.14786
	female	208	3.7391	1.12417	.16575
I wish to purchase environment friendly nature product that do not harm the environment and society.	male	196	3.6897	.82093	.10779
	female	208	3.9565	.84213	.12416
I can purchase the green product if the packaging & labeling information provided is sufficing my clearly doubts.	male	196	3.6724	1.04944	.13780
	female	208	3.5435	1.20566	.17777
I am in favor of the green product that any stages have used the sustainability process from renewable success.	male	196	3.1724	1.04526	.13725
	female	208	3.4783	1.24256	.18321
Quality of green product is Good.	male	196	3.0690	1.02362	.13441
	female	208	3.0000	1.19257	.17583
Natural ingredients as claimed	male	196	3.4828	1.27378	.16725
	female	208	3.0000	1.41421	.20851
Keeps promise of being safe for consumption	male	196	3.2069	1.21046	.15894
	female	208	3.1739	1.37120	.20217
Does not harm health	male	196	4.0517	.80399	.10557
	female	208	3.9348	1.06254	.15666
Perceived as good value for money preposition even if is priced little higher	male	196	3.9655	1.21346	.15933
	female	208	4.0435	1.07407	.15836
Advantage outweigh the disadvantage so can pay more for sake of owes and society	male	196	3.1379	.86751	.11391
	female	208	3.6087	1.02151	.15061
I take conscious decision of buying the greener attentive product for the sake of my duty towards the environment.	male	196	3.4828	.62804	.08247
	female	208	3.6087	.64904	.09570

Environment-Friendly Products—Adapt Green Now

I help others by persuading them to be environmentally cautious citizens.	male	196	2.7241	1.00513	.13198
	female	208	3.1739	1.12159	.16537
I believe that my simple/Small action can cause big impact towards the eco-system balance.	male	196	2.9310	1.22635	.16103
	female	208	3.0870	1.34703	.19861
I am inspired by the other who works toward having a automatable society	male	196	3.3621	.83136	.10916
	female	208	3.3261	1.19358	.17598
I believe in the individual empowerment necessary for it.	male	196	3.5345	.77721	.10205
	female	208	3.9783	.90650	.13366
In unrelieved to buy environmentally safe products for my personal happiness and satisfaction.	male	196	3.8966	.89226	.11716
	female	208	4.0435	.81531	.12021
It improves my quality of wellbeing in life.	male	196	3.2586	.73890	.09702
	female	208	3.8043	.98024	.14453
It gives a trek message to my peer group	male	196	2.8103	.90722	.11912
	female	208	2.9783	1.14483	.16880
It symbolizes the culture and values I possess	male	196	3.3276	1.34279	.17632
	female	208	3.7174	1.25898	.18563
It shone my deep rootedness to cultural/historical roots.	male	196	2.2759	1.32179	.17356
	female	208	2.8696	1.43927	.21221
It gives me more acceptances in my primary/Secondary /treasury groups	male	196	2.3276	1.40660	.18470
	female	208	3.0652	1.58328	.23344
I feel pride in explaining eh pros of eco-product, its information and how I am doing my contribution to society at larger	male	196	2.9655	1.10764	.14544
	female	208	3.1304	1.37612	.20290
I respect the wish of my other us my social group.	male	196	3.1897	.98153	.12888
	female	208	3.1087	1.17810	.17370
I support the decisions taken during purchase by any peer-group/Family	male	196	3.7241	.95133	.12492
	female	208	3.7391	.95300	.14051
The problem of depleting natural resources can be resolved by my message to society for green.	male	196	3.2414	.84418	.11085
	female	208	3.5652	1.00338	.14794
I am worried about the stale of affairs in an environment and what holds in the future	male	196	4.0000	.89834	.11796
	female	208	3.5652	1.24101	.18298
I think my initiative can lessen the severely abused/upset evtal balance	male	196	3.2069	1.08835	.14291
	female	208	3.5217	1.04858	.15460
Humans must live in harmony with the eco-systems/nature in order to survive.	male	196	3.6034	.99012	.13001
	female	208	3.8043	1.00265	.14783

I make special effort to reduce the use of paper non-xcylable product to pass on this message to the society.	male	196	3.5000	1.15849	.15212
	female	208	3.5870	1.00169	.14769
I don't mind spending on eco/green product if it convinces me on consistency and trustworthiness of brand being herbal	male	196	3.0690	.81353	.10682
	female	208	3.3478	1.33695	.19712
I can represent my socio-economic status before my poual groups by purchasing green	male	196	3.4138	1.15522	.15169
	female	208	3.7826	1.20946	.17833
I am competent to pay more	male	196	2.7069	.89851	.11798
	female	208	2.7609	1.17728	.17358
It is my moral obligation to pay more for environment	male	196	2.5172	.92227	.12110
	female	208	2.5652	1.12846	.16638
My culture/ Value makes me pay more for environment	male	196	1.9310	1.22635	.16103
	female	208	2.7826	1.50426	.22179
I am status conscious	male	196	1.8448	1.05634	.13870
	female	208	2.5435	1.53053	.22566
I can catalyze sustainable environment by paying and enchase to the engagement & involvement & Encouragement in working as a catalyst to enable green product advantage ones non-green and hence wi	male	196	2.7069	1.16992	.15362
	female	208	2.7609	1.31968	.19458
Trustworthiness of marketer's proofs of being eco-friendly	male	196	3.3276	1.04944	.13780
	female	208	3.4565	1.12953	.16654
Price as major factor in slow growth of sales of green products.	male	196	4.0690	1.05734	.13884
	female	208	3.8261	1.35490	.19977
Transparency of organization using green awareness and acceptance marketing concept since long.	male	196	3.7069	.72568	.09529
	female	208	4.0217	.80247	.11832
Employees/Customer feel that routine can be affected by implementing green concept.	male	196	4.0345	.95450	.12533
	female	208	3.9565	.89335	.13172
It is difficult for all companies to adapt to green initiatives compliantly	male	196	3.8103	.99924	.13121
	female	208	3.9783	.95427	.14070
Transparency n contents of the product labeled green.	male	196	4.0690	1.09002	.14313
	female	208	3.6522	1.28612	.18963

Table 4.7.2: Independent Samples Test

		Levene's Test for Equality of Variances		t-test for Equality of Means						
		F	Sig.	t	df	Sig. (2-tailed)	Mean Difference	Standard Error Difference	95% Confidence Interval of the Difference	
									Lower	Upper
I think green process of production respects the environmental health	Equal variances assumed	2.612	.109	.202	402	.840	.05397	.26689	-.47540	.58335
	Equal variances not assumed			.198	87.488	.843	.05397	.27246	-.48753	.59548
I think that the green product does not carry harmful ingredients like insecticide/pesticide/non-biodegradable or non-recyclable input materials	Equal variances assumed	1.756	.188	1.496	402	.138	.35682	.23854	-.11631	.82996
	Equal variances not assumed			1.479	91.937	.143	.35682	.24127	-.12237	.83601
I think that environment friendly products may cause quite less harm our environment as compared to its chemically laded counterpart	Equal variances assumed	2.045	.156	-1.540	402	.127	-.25487	.16555	-.58325	.07350
	Equal variances not assumed			-1.584	101.982	.116	-.25487	.16094	-.57410	.06436

I believe that with advent of latest technologies, use of eco-friendly products will increase and improve the efficiency in reduce-recycle reuse chain to support greener environment.	Equal variances assumed	2.052	.155	.397	402	.692	.10795	.27197	-.43150	.64739
	Equal variances not assumed			.391	90.474	.697	.10795	.27595	-.44024	.65614
The seal of quality organizations will further support & increase purchase of Eco- for greener environment.	Equal variances assumed	.674	.414	.054	402	.957	.00900	.16799	-.32421	.34220
	Equal variances not assumed			.054	101.200	.957	.00900	.16519	-.31869	.33668
I think that it is ethical on my part if I support the green products because they have (i) well defined origin (ii) are natural & safe foe consumption and in disposal.	Equal variances assumed	.283	.596	-1.021	402	.310	-.17691	.17329	-.52064	.16682
	Equal variances not assumed			-1.014	93.932	.313	-.17691	.17449	-.52338	.16955
There is respect for our nature in doing so.	Equal variances assumed	2.801	.097	-2.297	402	.224	-.41679	.18146	-.77671	-.05687
	Equal variances not assumed			-2.320	9775	.022	-.41679	.17967	-.77326	-.06032

My ethics, Values, norms and culture motivates me to stretch a little to be ethical toward the environment I live in.	Equal variances assumed	1.802	.182	-1.602	402	.112	-.21289	.13291	-.47651	.05073
	Equal variances not assumed			-1.595	95.008	.114	-.21289	.13348	-.47789	.05210
I am interested purchasing eco whose by product are used for renewable initiative	Equal variances assumed	.127	.723	-.276	402	.783	-.06897	.25014	-.56511	.42718
	Equal variances not assumed			-.274	94.558	.785	-.06897	.25150	-.56828	.43035
The purchase of green product is for the reason that I want to do may bit for our environment.	Equal variances assumed	.305	.582	-2.551	402	.612	-.56672	.22216	-1.00737	-.12607
	Equal variances not assumed			-2.551	96.748	.012	-.56672	.22212	-1.00757	-.12586
I wish to purchase environment friendly nature product that do not harm the environment and society.	Equal variances assumed	.114	.736	-1.628	402	.107	-.26687	.16394	-.59204	.05831
	Equal variances not assumed			-1.623	95.546	.108	-.26687	.16443	-.59327	.05954
I can purchase the green product if the packaging & labeling information provided is sufficing my clearly doubts.	Equal variances assumed	2.353	.128	.583	402	.561	.12894	.22133	-.31008	.56795
	Equal variances not assumed			.573	8746	.568	.12894	.22492	-.31792	.57580

I am in favor of the green product that any stages have used the sustainability process from renewable success.	Equal variances assumed	3.153	.079	-1.363	402	.176	-.30585	.22439	-.75093	.13923
	Equal variances not assumed			-1.336	87.841	.185	-.30585	.22891	-.76078	.14908
Quality of green product is Good.	Equal variances assumed	.813	.369	.317	402	.752	.06897	.21745	-.36234	.50027
	Equal variances not assumed			.312	88.970	.756	.06897	.22132	-.37080	.50873
Natural ingredients as claimed	Equal variances assumed	.828	.365	1.828	402	.070	.48276	.26408	-.04104	1.00656
	Equal variances not assumed			1.806	91.599	.074	.48276	.26731	-.04817	1.01368
Keeps promise of being safe for consumption	Equal variances assumed	.762	.385	.130	402	.897	.03298	.25348	-.46979	.53576
	Equal variances not assumed			.128	90.517	.898	.03298	.25717	-.47789	.54386
Does not harm health	Equal variances assumed	2.211	.140	.639	402	.524	.11694	.18302	-.24608	.47996
	Equal variances not assumed			.619	81.827	.538	.11694	.18891	-.25888	.49276
Perceived as good value for money preposition even if is priced little higher	Equal variances assumed	.729	.395	-.342	402	.733	-.07796	.22785	-.52990	.37397
	Equal variances not assumed			-.347	100.729	.729	-.07796	.22465	-.52361	.36769
Advantage outweigh the disadvantage so can pay more for sake of owes and society	Equal variances assumed	6.601	.012	-2.540	402	.113	-.47076	.18531	-.83832	-.10321
	Equal variances not assumed			-2.493	88.375	.015	-.47076	.18884	-.84602	-.09551

ENVIRONMENT-FRIENDLY PRODUCTS— ADAPT GREEN NOW

I take conscious decision of buying the greener attentive product for the sake of my duty towards the environment.	Equal variances assumed	.004	.952	-1.001	402	.319	-.12594	.12584	-.37555	.12367
	Equal variances not assumed			-.997	95.202	.321	-.12594	.12633	-.37672	.12485
I help others by persuading them to be environmentally cautious citizens.	Equal variances assumed	.422	.517	-2.153	402	.314	-.44978	.20890	-.86414	-.03541
	Equal variances not assumed			-2.126	91.330	.036	-.44978	.21158	-.87003	-.02952
I believe that my simple/ Small action can cause big impact towards the eco-system balance.	Equal variances assumed	.396	.531	-.617	402	.539	-.15592	.25291	-.65757	.34573
	Equal variances not assumed			-.610	92.166	.543	-.15592	.25569	-.66372	.35188
I am inspired by the other who works toward having a automatable society	Equal variances assumed	6.080	.015	.181	402	.857	.03598	.19889	-.35851	.43047
	Equal variances not assumed			.174	77.262	.863	.03598	.20709	-.37637	.44833
I believe in the individual empowerment necessary for it.	Equal variances assumed	1.226	.271	-2.686	402	.018	-.44378	.16520	-.77144	-.11611
	Equal variances not assumed			-2.639	88.908	.010	-.44378	.16816	-.77792	-.10964
In unrelieved to buy environmentally safe products for my personal happiness and satisfaction.	Equal variances assumed	.002	.961	-.866	402	.388	-.14693	.16963	-.48338	.18953
	Equal variances not assumed			-.875	9918	.384	-.14693	.16786	-.47996	.18611

It improves my quality of wellbeing in life.	Equal variances assumed	11.337	.001	-3.237	402	.012	-.54573	.16857	-.88009	-.21136
	Equal variances not assumed			-3.135	81.612	.022	-.54573	.17407	-.89204	-.19941
It gives a tiek message to my peer group	Equal variances assumed	1.306	.256	-.835	402	.406	-.16792	.20117	-.56693	.23110
	Equal variances not assumed			-.813	84.449	.419	-.16792	.20660	-.57873	.24289
It symbolizes the culture and values I possess	Equal variances assumed	.811	.370	-1.511	402	.134	-.38981	.25794	-.90144	.12183
	Equal variances not assumed			-1.523	9127	.131	-.38981	.25602	-.89779	.11818
It shows my deep rootedness to cultural/ historical roots.	Equal variances assumed	.024	.878	-2.187	402	.130	-.59370	.27144	-1.13211	-.05529
	Equal variances not assumed			-2.166	92.620	.033	-.59370	.27414	-1.13813	-.04928
It gives me more acceptances in my primary/ Secondary / treasury groups	Equal variances assumed	1.485	.226	-2.512	402	.016	-.73763	.29361	-1.32001	-.15525
	Equal variances not assumed			-2.478	90.862	.015	-.73763	.29767	-1.32893	-.14633
I feel pride in explaining eh pros of eco-product, its information and how I am doing my contribution to society at larger	Equal variances assumed	4.351	.039	-.677	402	.500	-.16492	.24350	-.64790	.31806
	Equal variances not assumed			-.661	85.337	.511	-.16492	.24964	-.66124	.33141

Environment-Friendly Products— Adapt Green Now

		F	Sig.	t	df	Sig. (2-tailed)	Mean Difference	Std. Error Difference	CI Lower	CI Upper
I respect the wish of my other us my social group.	Equal variances assumed	.990	.322	.382	402	.703	.08096	.21179	-.33912	.50104
	Equal variances not assumed			.374	87.297	.709	.08096	.21629	-.34893	.51084
I support the decisions taken during purchase by any peer-group/Family	Equal variances assumed	.007	.932	-.080	402	.937	-.01499	.18797	-.38783	.35785
	Equal variances not assumed			-.080	96.601	.937	-.01499	.18801	-.38816	.35817
The problem of depleting natural resources can be resolved by my message to society for green.	Equal variances assumed	6.318	.014	-1.787	402	.077	-.32384	.18121	-.68327	.03559
	Equal variances not assumed			-1.752	87.850	.083	-.32384	.18486	-.69122	.04354
I am worried about the environment and what holds in the future	Equal variances assumed	8.407	.005	2.071	402	.052	.43478	.20992	.01841	.85115
	Equal variances not assumed			1.997	7355	.049	.43478	.21770	.00149	.86808
I think my initiative can lessen the severely upset environmental balance	Equal variances assumed	.958	.330	-1.489	402	.140	-.31484	.21145	-.73425	.10457
	Equal variances not assumed			-1.495	98.170	.138	-.31484	.21054	-.73263	.10295
Humans must live in harmony with the eco-systems/nature in order to survive.	Equal variances assumed	.101	.752	-1.022	402	.309	-.20090	.19658	-.59081	.18902
	Equal variances not assumed			-1.020	96.129	.310	-.20090	.19687	-.59167	.18987

I make special effort to reduce the use of paper non-xcylable product to pass on this message to the society.	Equal variances assumed	1.059	.306	-.403	402	.688	-.08696	.21562	-.51463	.34072
	Equal variances not assumed			-.410	101.203	.683	-.08696	.21202	-.50754	.33362
I don't mind spending on eco/green product if it convinces me on consistency and trustworthiness of brand being herbal	Equal variances assumed	1193	.000	-1.312	402	.192	-.27886	.21250	-.70035	.14263
	Equal variances not assumed			-1.244	70.510	.218	-.27886	.22421	-.72597	.16825
I can represent my socio-economic status before my poual groups by purchasing green	Equal variances assumed	.113	.738	-1.584	402	.116	-.36882	.23287	-.83071	.09307
	Equal variances not assumed			-1.575	94.586	.119	-.36882	.23411	-.83362	.09599
I am competent to pay more	Equal variances assumed	4.302	.041	-.265	402	.791	-.05397	.20352	-.45766	.34971
	Equal variances not assumed			-.257	82.313	.798	-.05397	.20988	-.47147	.36352
It is my ethical obligation to pay more for environment	Equal variances assumed	4.960	.028	-.239	402	.812	-.04798	.20107	-.44679	.35084
	Equal variances not assumed			-.233	86.206	.816	-.04798	.20579	-.45705	.36110
My culture/ Value makes me pay more for environment	Equal variances assumed	2.254	.136	-3.181	402	.042	-.85157	.26772	-1.38260	-.32055
	Equal variances not assumed			-3.107	86.065	.003	-.85157	.27408	-1.39643	-.30672

Environment-Friendly Products— Adapt Green Now

I am status conscious	Equal variances assumed	839	.002	-2.749	402	.091	-.69865	.25415	-1.20275	-.19455
	Equal variances not assumed			-2.638	76.773	.010	-.69865	.26488	-1.22612	-.17118
I can catalyze sustainable environment by paying and enchase to the engagement & involvement & Encouragement in working as a catalyst to enable green product advantage ones non-green and hence wi	Equal variances assumed	1.345	.249	-.221	402	.826	-.05397	.24447	-.53888	.43093
	Equal variances not assumed			-.218	90.747	.828	-.05397	.24791	-.54643	.43848
Trustworthiness of marketer's proofs of being eco-friendly	Equal variances assumed	.338	.563	-.602	402	.549	-.12894	.21432	-.55403	.29616
	Equal variances not assumed			-.596	93.215	.552	-.12894	.21616	-.55817	.30030
Price as major factor in slow growth of sales of green products.	Equal variances assumed	2.881	.093	1.027	402	.307	.24288	.23648	-.22618	.71194
	Equal variances not assumed			.998	83.575	.321	.24288	.24328	-.24094	.72669
Transparency of organization using green awareness and acceptance marketing concept since long.	Equal variances assumed	.247	.620	-2.097	402	.338	-.31484	.15015	-.61267	-.01702
	Equal variances not assumed			-2.072	91.812	.041	-.31484	.15192	-.61657	-.01311

Statement		F	Sig.	t	df	Sig.	Mean Diff	Std Error	Lower	Upper
Employees/ Customer feel that routine can be affected by implementing green concept.	Equal variances assumed	.248	.620	.425	402	.671	.07796	.18322	-.28546	.44138
	Equal variances not assumed			.429	9185	.669	.07796	.18182	-.28279	.43872
It is difficult for all companies to adapt to green initiatives compliantly	Equal variances assumed	.012	.911	-.868	402	.387	-.16792	.19342	-.55156	.21573
	Equal variances not assumed			-.873	98.493	.385	-.16792	.19238	-.54967	.21384
Transparency n contents of the product labeled green.	Equal variances assumed	1.191	.278	1.788	402	.077	.41679	.23308	-.04553	.87911
	Equal variances not assumed			1.754	88.262	.083	.41679	.23758	-.05533	.88891

The results of Independent t-test shows that apart from certain aspect such as 'There is respect for our nature in doing so.', 'the purchase of green product is for the reason that I want to do may bit for our environment.', 'Advantage outweigh the disadvantage so can pay more for sake of owes and society', 'I help others by persuading them to be environmentally cautious citizens', 'I believe in the individual empowerment necessary for it.', 'It improves my quality of wellbeing in life.', 'It shows my deep rootedness to cultural/historical roots.', 'It gives me more acceptances in my primary/Secondary /treasury groups', 'I am worried about the environment and what holds in the future', 'My culture/ Value makes me pay more for environment', 'I am status conscious', 'Transparency of organization using green awareness and acceptance marketing concept since long.' Male and Female consumers do not differ on other aspect of the Perception about Environment Friendly Products.

4.8 Extent of Purchase of Environment Friendly FMCG Products

The researchers form table 4.6 wanted to find out how often does the respondents(total 404) purchase that selected FMCG products, with the aim to find out the frequency of purchase for green products in order to check the lifestyle change which do not exploit the environment. It was also found out that more the 'Green' product purchase, does it lead to greatly reduced harmful impact on our eco-system as compared to the products that have been manufactured, packaged and marketed in a conventional manner the extent of purchase of an environment friendly FMCG product has an impaction image carried by all stakeholders of the organization in a the way and if is also revealing that the company is seen as more socially responsible and value- oriented.

In the table 4.8, the researchers calculated the mean responses for each statement. The results reveled that Juices, Health Drinks, Snack Foods, Bottled water and vegetables tops the list in FMCG.

Table 4.8: FMCG Purchase					
	N	Min.	Max.	Mean	SD
How often do you purchase following Environmental friendly Products?- Juice	404	1	5	3.9231	1.21233
How often do you purchase following Environmental friendly Products?- Health Drinks	404	1	5	3.5385	1.53233

How often do you purchase following Environmental friendly Products?- Snack Food	404	1	5	3.3846	1.11745
How often do you purchase following Environmental friendly Products?- Bottled Water	404	1	5	3.3077	1.49483
How often do you purchase following Environmental friendly Products?- Vegetables	404	1	5	3.3077	1.2701
How often do you purchase following Environmental friendly Products?- Tea/ Coffee	404	1	5	3.1923	1.21541
How often do you purchase following Environmental friendly Products?- staples/ Cereals/ Spices	404	1	5	3.1923	1.18303
How often do you purchase following Environmental friendly Products?- Skin Care/ Hair Care	404	1	5	3.1923	1.52695
How often do you purchase following Environmental friendly Products?- Bakery	404	1	5	3	1.2148
How often do you purchase following Environmental friendly Products?- Cosmetics	404	1	5	2.8462	1.20492
How often do you purchase following Environmental friendly Products?- Personal Wash	404	1	5	2.8077	1.42159
How often do you purchase following Environmental friendly Products?- Air Freshner	404	1	5	2.8077	1.21541

How often do you purchase following Environmental friendly Products?- Hygiene-oral Care	404	1	5	2.7308	1.20057
How often do you purchase following Environmental friendly Products?- Mosquito repellent	104	1	5	2.7308	1.32365
Valid N (listwise)	404				

Fig 4.8: FMCG Purchase

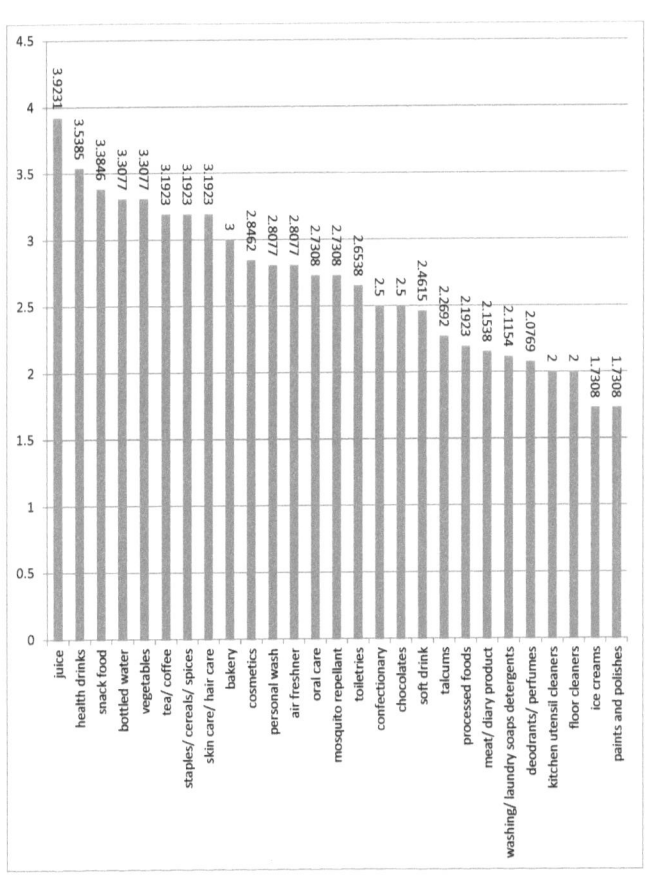

4.9 Extent of Purchase of Environment Friendly Textile Products

Textile(Extent of purchase) from the descriptive analysis in table 4.9, the researcher has tried to find out through a series of fourteen(14) statements, depicting the fourteen different type of environmentally friendly textile products for males, females, kids, youth, home-care, home furnishings, bath-care, kitchen-care and upholstery categories.

The objective had been to find out the category of textile product being bought by the customers as to what extent to the environmental- friendliness of these products at various levels from the raw materials to the furnished goods and their final disposal as well.

From the table 4.9 one can evidently see how close are the individual data values of SD placed from its mean value, thus going also shape of the distribution of data values from the mean and how far a sample mean is placed from the population mean, i.e. the accuracy of individual sample mean with relation to the mean. The overall results convey that In textile, the wearables and dining mats/napkins dominate the others.

Table 4.9: Textile Purchase					
	N	Min.	Max.	Mean	SD
How often do you purchase following Environmental friendly Products?- suite/ kurta	404	1	5	2.7692	1.15943
How often do you purchase following Environmental friendly Products?- napkins	404	1	5	2.4231	1.42578

Environment-Friendly Products— Adapt Green Now

How often do you purchase following Environmental friendly Products?- lowers	404	1	5	2.3846	1.21725
How often do you purchase following Environmental friendly Products?- dining mats/napkins	404	1	5	2.3846	1.3388
How often do you purchase following Environmental friendly Products?- shawls	404	1	5	2.1538	1.35653
How often do you purchase following Environmental friendly Products?- floor mats	404	1	5	2.1538	1.3276
How often do you purchase following Environmental friendly Products?- bedsheets/ beddings	404	1	5	2.0769	1.21233
How often do you purchase following Environmental friendly Products?- scarfs	404	1	5	2.0769	1.54977
How often do you purchase following Environmental friendly Products?- saree	404	1	4	2.0385	1.23007
How often do you purchase following Environmental friendly Products?- curtains	404	1	5	1.9231	1.21233
How often do you purchase following Environmental friendly Products?- towels	404	1	5	1.8846	1.28704
How often do you purchase following Environmental friendly Products?- upholstery	404	1	4	1.8462	1.13864
How often do you purchase following Environmental friendly Products?- kitchen cloths	404	1	5	1.7692	1.22459

How often do you purchase following Environmental friendly Products?-tablecloths	404	1	5	1.5769	1.0492
Valid N (listwise)	404				

Fig. 4.9: Textile Purchase

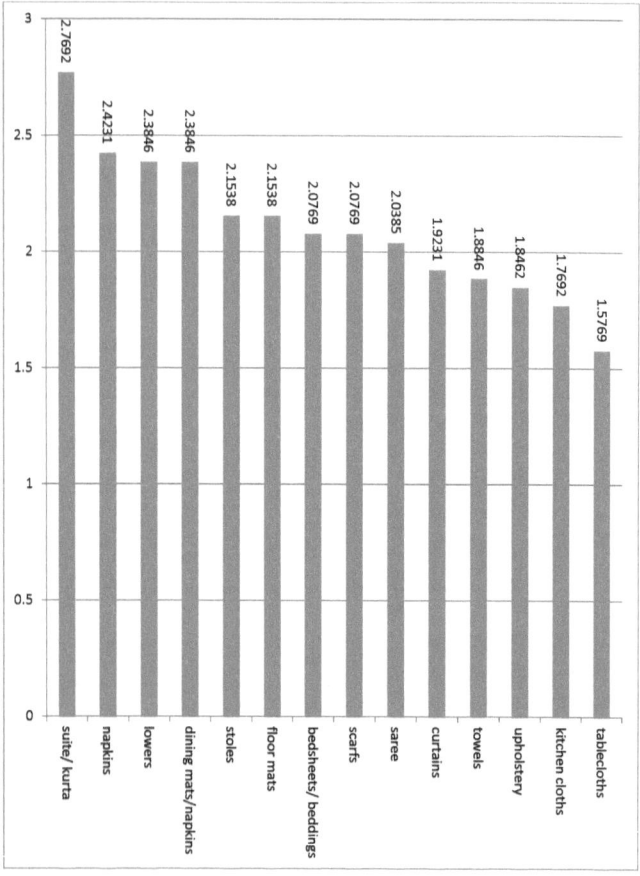

4.10 Factor Impacting the Green Products Purchase

To identify the various factors impacting the green products purchase decision the factor analysis was conducted. The following section presents the results of the same.

Table 4.10.1: KMO and Bartlett's Test		
Kaiser-Meyer-Olkin Measure of Sampling Adequacy.		0.944
Bartlett's Test of Sphericity	Approx. Chi-Square	15448.604
	df	1225
	Sig.	0

KMO and Barlett's Test indicates the suitability of the data for factor analysis. As we can observe, that the value of KMO test is 0.944 which indeed is greater than 0.50. Thus the above-mentioned value depicted from the test states that the data is reliable and the sample size is adequate in order to conduct factor analysis.

Table 4.10.2: Communalities		
	Initial	Extraction
I think green process of production respects the environmental health	1.000	.930
I think that the green product does not carry harmful ingredients like wisectiade/ pesticide/ plastic/non-biodegradable or non-recyclable input materials	1.000	.845
I think that environment friendly products may cause quite less harm our environment as compared to its chemically laded counterpart	1.000	.756

I believe that with advent of latest technologies, use of eco-friendly products will increase and improve the efficiency in reduce-recycle reuse chain to support greener environment.	1.000	.895
The seal of quality organizations will further support & increase purchase of Eco- for greener environment.	1.000	.655
I think that it is ethical on my part if I support the green products because they have (i) well defined origin (ii) are natural & safe foe consumption and in disposal.	1.000	.742
There is respect for our nature in doing so.	1.000	.680
My ethics, Values, norms and culture motivates me to stretch a little to be ethical toward the environment I live in.	1.000	.871
I am interested purchasing eco whose by product are used for renewable initiative	1.000	.824
the purchase of green product is for the reason that I want to do may bit for our environment.	1.000	.892
I wish to purchase environment friendly nature product that do not harm the environment and society.	1.000	.920
I can purchase the green product if the packaging & labeling information provided is sufficing my clearly doubts.	1.000	.772
I am in favor of the green product that any stages have used the sustainability process from renewable success.	1.000	.747
Quality of green product is Good.	1.000	.833
Natural ingredients as claimed	1.000	.901
Keeps promise of being safe for consumption	1.000	.865
Does not harm health	1.000	.941
Perceived as good value for money preposition even if is priced little higher	1.000	.931
Advantage outweigh the disadvantage so can pay more for sake of owes and society	1.000	.698

I take conscious decision of buying the greener attentive product for the sake of my duty towards the environment.	1.000	.717
I help others by persuading them to be environmentally cautious citizens.	1.000	.907
I believe that my simple/Small action can cause big impact towards the eco-system balance.	1.000	.891
I am inspired by the other who works toward having a automatable society	1.000	.897
I believe in the individual empowerment necessary for it.	1.000	.862
In unrelieved to buy environmentally safe products for my personal happiness and satisfaction.	1.000	.829
It improves my quality of wellbeing in life.	1.000	.770
It gives a trek message to my peer group	1.000	.916
It symbolizes the culture and values I possess	1.000	.873
It shone my deep rootedness to cultural/historical roots.	1.000	.921
It gives me more acceptances in my primary/Secondary /treasury groups	1.000	.916
I feel pride in explaining eh pros of eco-product, its information and how I am doing my contribution to society at larger	1.000	.827
I respect the wish of my other us my social group.	1.000	.841
I support the decisions taken during purchase by any peer-group/Family	1.000	.804
The problem of depleting natural resources can be resolved by my message to society for green.	1.000	.658
I am worried about the environment and what holds in the future	1.000	.743
I think my initiative can lessen the severely upset environmental balance	1.000	.885

Humans must live in harmony with the eco-systems/nature in order to survive.	1.000	.913
I make special effort to reduce the use of paper non-recyclable product to pass on this message to the society.	1.000	.750
I don't mind spending on eco/green product if it convinces me on consistency and trustworthiness of brand being herbal	1.000	.801
I can represent my socio-economic status before my peer groups by purchasing green	1.000	.795
I am competent to pay more	1.000	.827
It is my ethical obligation to pay more for environment	1.000	.558
My culture/ Value makes me pay more for environment	1.000	.703
I am status conscious	1.000	.591
I can catalyze sustainable environment by paying and enchase to the engagement & involvement & Encouragement in working as a catalyst to enable green product advantage ones non-green and hence wi	1.000	.602
Trustworthiness of marketer's proofs of being eco-friendly	1.000	.752
Price as major factor in slow growth of sales of green products.	1.000	.780
Transparency of organization using green awareness and acceptance marketing concept since long.	1.000	.831
Employees/Customer feel that routine can be affected by implementing green concept.	1.000	.826
It is difficult for all companies to adapt to green initiatives compliantly	1.000	.734
Transparency n contents of the product labeled green.	1.000	.816
Extraction Method: Principal Component Analysis.		

The commonalities table shows that h² values for each statement are greater than .5, hence it further shows that the sample is appropriate for factor analysis and there is significant commonality among the various statements.

Table 4.10.3: Total Variance Explained

Component	Initial Eigenvalues			Extraction Sums of Squared Loadings			Rotation Sums of Squared Loadings		
	Total	% of Variance	Cumulative %	Total	% of Variance	Cumulative %	Total	% of Variance	Cumulative %
1	12.793	25.083	25.083	12.793	25.083	25.083	11.522	22.593	22.593
2	11.941	23.414	48.497	11.941	23.414	48.497	11.210	21.980	44.573
3	8.314	16.303	64.800	8.314	16.303	64.800	9.712	19.043	63.616
4	4.922	9.651	74.450	4.922	9.651	74.450	4.451	8.728	72.344
5	3.064	6.008	80.458	3.064	6.008	80.458	4.139	8.115	80.458
6	2.465	4.834	85.292						
7	1.796	3.522	88.814						
8	1.499	2.940	91.755						
9	1.368	2.683	94.437						
10	1.222	2.396	96.833						
11	.682	1.337	98.170						
12	.549	1.076	99.247						
13	.384	.753	100.000						
14	3.165E-015	6.207E-015	100.000						
15	2.771E-015	5.432E-015	100.000						
16	2.285E-015	4.481E-015	100.000						
17	2.167E-015	4.248E-015	100.000						
18	1.921E-015	3.767E-015	100.000						
19	1.834E-015	3.596E-015	100.000						
20	1.669E-015	3.272E-015	100.000						
21	1.375E-015	2.697E-015	100.000						
22	1.192E-015	2.337E-015	100.000						
23	1.173E-015	2.301E-015	100.000						
24	1.007E-015	1.975E-015	100.000						
25	9.366E-016	1.836E-015	100.000						
26	6.876E-016	1.348E-015	100.000						
27	5.143E-016	1.009E-015	100.000						
28	4.843E-016	9.497E-016	100.000						
29	3.951E-016	7.747E-016	100.000						
30	2.677E-016	5.249E-016	100.000						
31	8.399E-017	1.647E-016	100.000						
32	-3.758E-017	-7.370E-017	100.000						
33	-6.358E-017	-1.247E-016	100.000						
34	-2.427E-016	-4.759E-016	100.000						
35	-2.759E-016	-5.410E-016	100.000						
36	-3.747E-016	-7.348E-016	100.000						
37	-5.328E-016	-1.045E-015	100.000						
38	-6.235E-016	-1.223E-015	100.000						

39	-7.391E-016	-1.449E-015	100.000					
40	-7.856E-016	-1.540E-015	100.000					
41	-9.133E-016	-1.791E-015	100.000					
42	-1.189E-015	-2.331E-015	100.000					
43	-1.222E-015	-2.397E-015	100.000					
44	-1.302E-015	-2.552E-015	100.000					
45	-1.497E-015	-2.935E-015	100.000					
46	-1.803E-015	-3.535E-015	100.000					
47	-2.027E-015	-3.975E-015	100.000					
48	-2.150E-015	-4.216E-015	100.000					
49	-2.403E-015	-4.711E-015	100.000					
50	-2.942E-015	-5.769E-015	100.000					
51	-3.076E-015	-6.032E-015	100.000					
Extraction Method: Principal Component Analysis.								

From the table of total variance explained, one can analyze the percentage of component responsible for the change in variance. Here we can notice that component 1 accounts for 22.593%, component 2 for 21.980%, component 3 for 19.043%, component 4 for 08.728%, and component 5 for 08.115% of the variance.

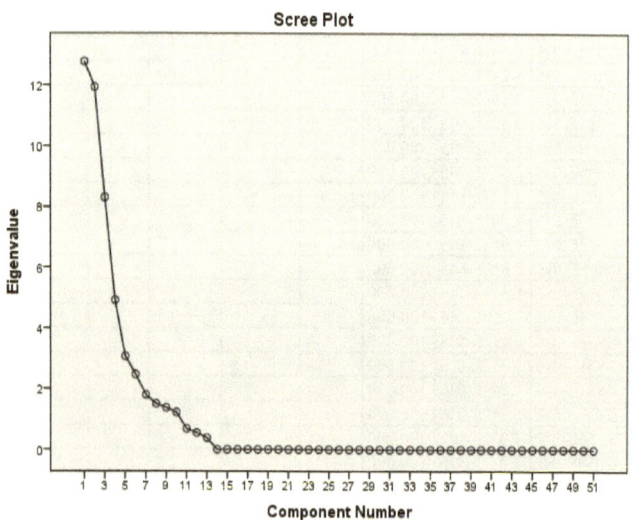

Table 4.10.4: Component Transformation Matrix					
Component	1	2	3	4	5
1	.623	.454	.616	.123	.107
2	-.650	.752	.104	-.012	.011
3	.416	.453	-.724	-.276	.147
4	-.110	-.148	.237	-.673	.676
5	-.064	-.032	-.169	.676	.714
Extraction Method: Principal Component Analysis. Rotation Method: Varimax with Kaiser Normalization.					

The component transformation matrix indicates the factor reduction. As we had 50 statements, with the help of factor analysis we had reduced the factors and it can be spotted in the above mentioned table where 5 factors are considered to be of utmost importance affecting the purchase of green products.

Table 4.10.5: Rotated Component Matrixa					
	Component				
	1	2	3	4	5
I think that the green product does not carry harmful ingredients like wisectiade/ pesticide/ plastic/non-biodegradable or non-recyclable input materials	-.644				
I think that it is ethical on my part if I support the green products because they have (i) well defined origin (ii) are natural & safe foe consumption and in disposal.	.804				
There is respect for our nature in doing so.	.761				

Natural ingredients as claimed	.544				
The purchase of green product is for the reason that I want to do may bit for our environment.	.914				
I help others by persuading them to be environmentally cautious citizens.	.805				
I believe in the individual empowerment necessary for it.	.808				
It symbolizes the culture and values I possess	.878				
It shone my deep rootedness to cultural/historical roots.	.839				
It gives me more acceptances in my primary/ Secondary /treasury groups	.856				
I think my initiative can lessen the severely upset environmental balance	.777				
Employees/Customer feels that routine can be affected by implementing green concept.	.511				
My culture/ Value makes me pay more for environment	.687				
I am status conscious	.554				
I can catalyze sustainable environment by paying and enchase to the engagement & involvement & Encouragement in working as a catalyst to enable green product advantage ones non-green		.684			

Environment-Friendly Products— Adapt Green Now

It is my obligation to pay more for environment	.531			
I feel pride in explaining eh pros of eco-product, its information and how I am doing my contribution to society at larger	.869			
I respect the wish of my other us my social group.	.899			
It gives a trek message to my peer group	.629			
Quality of green product is Good.	.874			
Keeps promise of being safe for consumption	.897			
I think green process of production respects the environmental health	.923			
I believe that with advent of latest technologies, use of eco-friendly products will increase and improve the efficiency in reduce-recycle reuse chain to support greener environment.	.869			
I am interested purchasing eco whose by product are used for renewable initiative	.818			
I wish to purchase environment friendly nature product that do not harm the environment and society.	.672			
I believe that my simple/ Small action can cause big impact towards the eco-system balance.	.533			

I am inspired by the other who works toward having a automatable society		.716			
I am worried about the environment and what holds in the future		.508			
I don't mind spending on eco/green product if it convinces me on consistency and trustworthiness of brand being herbal		.838			
Price as major factor in slow growth of sales of green products.		.575			
Transparency n contents of the product labeled green.		.648			
Perceived as good value for money preposition even if is priced little higher			.878		
Advantage outweigh the disadvantage so can pay more for sake of owes and society			.654		
I take conscious decision of buying the greener attentive product for the sake of my duty towards the environment.			.723		
It improves my quality of wellbeing in life.			.639		
I support the decisions taken during purchase by any peer-group/Family			.725		
The problem of depleting natural resources can be resolved by my message to society for green.			.638		

Humans must live in harmony with the eco-systems/nature in order to survive.			.851		
I make special effort to reduce the use of paper non-xcylable product to pass on this message to the society.			.761		
I can represent my socio-economic status before my poual groups by purchasing green			.856		
Trustworthiness of marketer's proofs of being eco-friendly			.808		
Transparency of organization using green awareness and acceptance marketing concept since long.			.887		
It is difficult for all companies to adapt to green initiatives compliantly			.648		
I am in favor of the green product that any stages have used the sustainability process from renewable success.				.722	
Does not harm health				.962	
I am competent to pay more				.843	
I think that environment friendly products may cause quite less harm our environment as compared to its chemically laded counterpart				.623	

The seal of quality organizations will further support & increase purchase of Eco- for greener environment.					.664
My ethics, Values, norms and culture motivates me to stretch a little to be ethical toward the environment I live in.					.672
I can purchase the green product if the packaging & labeling information provided is sufficing my clearly doubts.					.653
In unrelieved to buy environmentally safe products for my personal happiness and satisfaction.					.750
Extraction Method: Principal Component Analysis. Rotation Method: Varimax with Kaiser Normalization.					
a. Rotation converged in 17 iterations.					

On the basis of the rotated component matrix (Table 4.10.5), we have grouped several statements together in order to make it easier to analyze the content. From the above mentioned table we can observe that in each of the 5 factors there are some or the other value, which has the highest value in accordance to the specified statement. The different groups formed are named as: "Pro-Environmental Concern", "Social Group Influence", "Eco-Branding & Labelling", "Green Awareness and acceptance marketing Communication Tools", "Self-Image and Personal Values".

4.11 Impact of various factors on Green Product Purchase Decision

To find out the extent of impact of identified factors on the purchase of Eco-friendly/green products the multiple regression was conducted.

Model Summary

Model	R	R Square	Adjusted R Square	Standard Error of the Estimate
1	.923a	.853	.849	.48632

a. Predictors: (Constant), Factor1, Factor2, Factor3, Factor4, Factor5

ANOVAa

Model		Sum of Squares	df	Mean Square	F	Sig.
1	Regression	180.545	4	60.182	254.459	.000b
	Residual	31.219	399	.237		
	Total	211.765	403			

a. Dependent Variable: Purchase of Green Products.

Multiple Regression analyses were conducted to examine the relationship between purchase of green products and various potential predictors explored in section 4.10. The multiple regression model with all five predictors produced R^2 = .853, which shows that there is 85% of impact of independent variables ("Pro-Environmental Concern", "Social Group Influence", "Eco-Branding & Labelling", "Green Awareness and acceptance marketing Communication Tools", "Self-Image and Personal Values") on dependent variable (Purchase of Green Products).

4.12 Difference among Conventional Awareness and acceptance marketing and Neuro-Awareness and acceptance marketing

To identify possible differences in terms of inviting attention, creating Interest, infusing desire and promoting purchase action among the consumers of conventional awareness and acceptance marketing and neuro-awareness and acceptance marketing method the independent t-test was conducted.

	Table 4.12.1 Group Statistics				
	Type of Awareness and acceptance marketing	N	Mean	Standard Deviation	Standard Error Mean
1. The products of display have caught my attention.	Conventional Awareness and acceptance marketing	148	2.2500	.66844	.09648
	Sensory_Neuro Awareness and acceptance marketing	256	4.8961	.35665	.01890
2. I got Interested in these products.	Conventional Awareness and acceptance marketing	148	2.7500	.66844	.09648
	Sensory_Neuro Awareness and acceptance marketing	256	4.8961	.35665	.01890
3. These eco-friendly products initiated desire in me for purchase	Conventional Awareness and acceptance marketing	148	2.7500	.66844	.09648
	Sensory_Neuro Awareness and acceptance marketing	256	4.8961	.35665	.01890
4. I am willing to purchase someof these products.	Conventional Awareness and acceptance marketing	148	2.2500	.83793	.12094
	Sensory_Neuro Awareness and acceptance marketing	256	4.8624	.56732	.03007

Table 4.12.2 Independent Samples Test

		Levene's Test for Equality of Variances		t-test for Equality of Means					95% Confidence Interval of the Difference	
		F	Sig.	t	df	Sig. (2-tailed)	Mean Difference	Standard Error Difference	Lower	Upper
1. The products of display have caught my attention.	Equal variances assumed	61.926	.000	-42.422	402	.000	-2.64607	.06238	-2.76869	-2.52345
	Equal variances not assumed			-26.914	50.667	.000	-2.64607	.09831	-2.84347	-2.44866
2. I got Interested in these products.	Equal variances assumed	61.926	.000	-34.406	402	.000	-2.14607	.06238	-2.26869	-2.02345
	Equal variances not assumed			-21.829	50.667	.000	-2.14607	.09831	-2.34347	-1.94866
3. These eco-friendly products initiated desire in me for purchase	Equal variances assumed	61.926	.000	-34.406	402	.000	-2.14607	.06238	-2.26869	-2.02345
	Equal variances not assumed			-21.829	50.667	.000	-2.14607	.09831	-2.34347	-1.94866
4. I am willing to purchase some of these products.	Equal variances assumed	43.250	.000	-28.071	402	.000	-2.61236	.09306	-2.79531	-2.42941
	Equal variances not assumed			-20.962	52.962	.000	-2.61236	.12463	-2.86233	-2.36239

Attention

It is found from the analysis (Table 4.12.2), that there is a significance difference among the consumers of conventional awareness and acceptance marketing and neuro-awareness and acceptance marketing method towards the attention they gained concerning the green products on display. When compared to a conventional awareness and acceptance marketing (Table 4.12.1), the neuro-awareness and acceptance marketing method gained more attention concerning the products on display. Hence, it could be concluded that green products marketed using the neuro-awareness and acceptance marketing method gets more consumer attention.

Interest

It is found from the analysis (Table 4.12.2), that there is a significance difference among the consumers of conventional awareness and acceptance marketing and neuro-awareness and acceptance marketing method towards the aspect 'Interest' concerning the green products on display. When compared to a conventional awareness and acceptance marketing (Table 4.12.1), the neuro-awareness and acceptance marketing method created more interest in the products on display. Hence, it could be concluded that if the green products are marketed using the neuro-awareness and acceptance marketing method, consumers get more interested in the products.

Desire

It is found from the analysis (Table 4.12.2), that there is a significance difference among the consumers of conventional awareness and acceptance marketing and neuro-awareness and acceptance marketing method towards the desire they infused concerning the green products on display. When compared to a (Table 4.12.1) conventional awareness and acceptance marketing, the neuro-awareness and acceptance marketing method infused more desire concerning the products on display. Hence it could be concluded that if the green products are marketed using the neuro-awareness and acceptance marketing method consumer's desire for the product purchase get stimulated.

Action

It is found from the analysis (Table 4.12.2), that there is a significance difference among the consumers of conventional awareness and acceptance marketing and neuro-awareness and acceptance marketing method towards the aspect 'action'. When compared to a conventional awareness and acceptance marketing(Table 4.12.1), the neuro-awareness and acceptance marketing consumers were more actionable in terms of purchase of the green products on display. Hence it could be concluded that if the products are marketed using the neuro-awareness and acceptance marketing method chances are that more consumers will be willing to purchase the marketed green products.

Based on the results, analysis and its interpretation an attempt has been made in the next chapter to discuss the findings of the present research work.

Chapter 4

DISCUSSION

The study explores green awareness and acceptance marketing as a phenomenon and the role of neuro-awareness and acceptance marketing of green products on the purchase decision. Main objectives of the study were to understand the consumer's perception towards Eco-friendly/green products; to study the extent of purchase of Eco-friendly/green products; to explore the factors affecting the purchase of Eco-friendly/green products; to find the extent of impact of identified factors on the purchase of Eco-friendly/green products and ultimately to identify possible differences in terms of inviting attention, creating Interest, infusing desire and promoting purchase action among the consumers of conventional awareness and acceptance marketing and neuro-awareness and acceptance marketing method. These objectives were examined by articles and literature, and also by primary data collection.

Results of the survey show that price, had been the most significant reason for non-purchase, and while lack of standardization of green products is another reason because of which green products are slow in market growth as compared to their non-green counterparts.

With this research, I examined the level of consumer willingness to pay more for eco-friendly products and consumers willing to recycle e-waste at drop-off recycling centers. The research process has been both rewarding and challenging for two reasons. First, the results of this research provided significant insights into the decision-making process customers go though and helped focus on which variables (price, quality, or brand name) play an important part in the final purchase decision. Secondly, the literature and research findings revealed consumers (Chang & Fong, 2010) and business (Eco-promising, 2008) views on eco-friendly products. I believe that the information from this study provides a point in time reference of participating customers' spending decisions and propensity to recycle. I was surprised that business managers have not noticed the eco-friendly product trend sooner since this information about scarcity of resources has been around for several years. Based upon completing this study, I believe as customer demand for ecofriendly products increases, business managers will respond and more eco-friendly products will be available in stores along with people using drop-off center to recycle goods.

Summary and Conclusions In conclusion, to promote green products as the wave of the future, the focus should be on product stewardship and product awareness and acceptance marketing. Because evolving and changing customers' views drive business product development, it is the customers' expressing their newly found interest in green products that should prompt businesses leaders to refocus their efforts and dedicate their resources to explore how they can 96 harness this new and potentially competitive advantage to increase companies' bottom lines while satisfying the customer base.

The impact of consumer perception (respondents) in buying environment friendly textile products is based on the image they carry about the shopping stores. It can be concluded that a single attribute seals focus on the product being eco-friendly and that attributed is government certification revealing trust, and loyalty and so they carry that security certified label with more trust worthiness. For FMCG products also it can be concluded that consumers prefer eco-purchase stores which they perceive to be of high Brand trust because of their mission to save the environment eco-system and visualize a chemical- free society at large.

The results findings descriptive statistics on knowledge and awareness about 'Green' 'Herbal' 'Eco' environmentally sustainable products in the textile and FMCG sectors shows that majority of the respondents, were in agreement with the various questions answered to test their knowledge and understanding about what they perceive to be having in a product which is labeled 'Green' or eco-friendly or causing minimum or no hazard to the environment and to nature.

According to the results about the respondent's perception on green products one would conclude that respondents feel that the seal of quality organizations further support & increase purchase of Eco- for greener environment. They think that environment friendly products may cause quite less harm our environment as compared to its chemically laded counterpart. It is because they have respect for our nature also their ethics, values, norms and culture motivates them to stretch a little to be ethical toward the environment they live in. However, they strongly feel that a green product does not harm the health. The results of Independent t-test shows that apart from certain aspect such as 'There is respect for our nature in doing so.', 'the purchase of green product is for the reason that I want to do may bit for our environment.', 'Advantage outweigh the disadvantage so can pay more for sake of owes and society', 'I help others by persuading them to be environmentally cautious citizens', 'I believe in the individual empowerment necessary for it.', 'It improves my quality of wellbeing in life.', 'It shows my deep rootedness to cultural/historical roots.', 'It gives me more acceptances in my primary/Secondary /treasury groups', 'I am worried about the environment and what holds in the future', 'My culture/ Value makes me pay more for environment', 'I am status conscious', 'Transparency of organization using green awareness and acceptance marketing concept since long.' Male and Female consumers do not differ on other aspect of the Perception about Environment Friendly Products. At last, the impulse to go green is spreading faster than a morning glory. Consumers too are getting behind the idea of being greener. In almost every opinion poll, consumers say that they are very concerned about climate change. They worry about rising seas, declining air quality, shrinking animal habitats, lengthening droughts, and newly brewing diseases. Even the green goods that have caught on have tiny market shares.

Consumers want to act green, but they expect businesses to lead the way. According to a global survey, 61 percent of consumers say that corporations should take the lead in tackling the issue of climate change. To do this, businesses need to develop more and better Earth-friendly products. Some already are, but they are not doing a good job of awareness and acceptance marketing them, finds a Climate Group study, which discovered that two-thirds consumers cannot name a green brand. Corporations can reap multiple benefits by going green. They can reduce their energy consumption, lessen their risks, meet competitive threats, enhance their brands, and increase their revenues. To realize the potential of the green market, businesses must help consumers change their behaviors. First, consumers have to be aware that a product exists before they buy it and consumers must believe that a product will get the job done. But many believe that green products are of lower quality than their more traditional "browner" counterparts. Consumers often believe that the prices of green goods are too high, and have a hard time finding them anyway.

To increase the sales of environmentally sensible products, companies must remove barriers, mainly, lack of awareness, negative perceptions, distrust, high prices, and low availability. In other words, they must increase consumers' awareness of green products, improve consumers' perceptions of eco-products' quality, strengthen consumers' trust, lower the prices of green products, and increase these products' availability.

> *An honest attempt has been made in the preceding chapter to answer all the objectives formulated in the research work and also present some of the observations of the researchers about the paradigm of neuro-awareness and acceptance marketing. In the next chapter an attempt has been made to present the Implications, Limitations and Recommendation for further list.*

Chapter 5

IMPLICATIONS, LIMITATIONS AND RECOMMENDATION FOR FURTHER RESEARCH

Implications

This thesis examined consumers' perceptions and attitudes towards eco-friendly products, their usage pattern and how the neuro-awareness and acceptance marketing influence the green products consumption. This study contributes to theory and practice in neuro awareness and acceptance marketing and green consumption, and argues that neuro awareness and acceptance marketing does influence the consumption of green products. The use of neuro-awareness and acceptance marketing in green products would assist with enhancing the social influences that consumers are engaging in. This reflection may enable marketers to develop proactive campaigns with the view that future generations will engage in green practices and be influenced to purchase green products.

There may be opportunities for firms to use neuro awareness and acceptance marketing in promoting green products.

Businesses alone cannot lead consumers from intention to action. In many instances, the government and the civil sector need to be heavily involved to achieve long-lasting changes in consumer behavior. Nevertheless, businesses should play a leading role in the green movement in order to shape their market opportunities and manage potential regulation of their industries.

Green products and services are only a niche market today, but they are poised for strong growth. Entering the green market can also improve companies' reputation, thereby increasing the value of their brands. In addition, firms that have a strong position in the green market can stay ahead of regulation and protect their market share from competitors. Companies may rightly ask whether cultivating green consumers is worth all the trouble. The study reveals that the belief development is imperative for success. Once businesses remove the obstacles between consumers' desire to buy green and the actual follow-through of those sentiments, green products could experience explosive sales growth. What's more, building a reputation as an Earth-friendly corporation can do much more than generate increased revenues from green products.

Limitation

Although this thesis contributes to neuro awareness and acceptance marketing and eco-friendly products theory and practice, the research design contains inherent limitations both directly relating to this research and generically to the methodological approach. Regarding the credibility of the findings, this study does not claim to represent the views of all consumers throughout India as the respondents were chosen randomly from the different places of the Delhi and NCR only; results from larger national and international studies might be different, wholly or in part. It is acknowledged that the views expressed were also subject to the researcher's interpretation, pessimistic or unconventional views may have been suppressed. However, many of the findings are consistent with, or build on the literature; thus, adding weight to their credibility. These findings are relevant to the specific context of neuro awareness and acceptance marketing, green consumption, and the time the research was collected. The study in this thesis does, however, provide a unique insight and important contributions to current research. The analysis presented here does lean towards several ways in which social practice theory might be integrated through further empirical applications of this kind.

Recommendation for Further Research

The research in this thesis provides several ideas for future research to inform and inspire research in neuro awareness and acceptance marketing and its uses for awareness and acceptance marketing eco-friendly products. Discussed in this section is the opportunity to replicate the research and the notion of different methodologies as ways of conducting quantitative research. The processes within this research were reported in detail in Chapter Three, thereby enabling a future researcher to repeat the work. Such in-depth coverage also allows the reader to assess the extent to which proper research practices were followed. This enables readers of the research thesis to develop a thorough understanding of the methods and their effectiveness. Consumers tend to blur the lines between consumption of green products and services and their everyday green practices. Therefore, other social practice could provide further interesting insights into this area of research.

As this study utilised quantitative data, it aimed to provide a detailed description of consumers' perceptions and attitudes. Therefore, its findings provide an informative framework for further qualitative research, which is recommended in order to validate and extend its findings. Additional qualitative research that takes the thematic analyses up towards theory building by developing analytical categories may assist to build conceptual models for future quantitative testing. Further research, both qualitative and quantitative, particularly the pursuit of more detailed studies in more settings, appears absolutely vital.

REFERENCES

- Aaker, A, Kumar, VD & George; Awareness and acceptance marketing research, John Wiley and Sons, Inc, New York.
- AB Hamid, NR & Kassim, N; 'Internet technology as a tool in managing customer relationships', The Journal of American Academy of Business Cambridge, vol. 4, no.1&2, pp.103-108, 2004.
- Ann, K. Amir, G. and Luc, W. (2012). "Go Green! Should Environmental Messages Be So Assertive?". *Journal of Awareness and acceptance marketing.* Vol 46, pp. 95-102.
- Anselmsson and Johansson (2007) corporate social responsibility and the positioning of grocery brands, *International Journal of Retail & Distribution Management*, Vol.35 No.10, pp. 835-866.
- Babin, B. J., & Babin, L. (2001). "Seeing something different: A model of schema typically, consumer affect, purchase intentions and perceived shopping value". *Journal of Business Research.* 54 pp. 89-96.

- Balderjahn, I. (1988). "Personality variables and environmental attitudes as predictors of ecologically responsible consumption patterns". *Journal of Business Research.* 17 pp. 51 –56.
- Burns, AC & Bush, RF; Awareness and acceptance marketing research, Prentice Hall International, Inc., New Jersey, 2000.
- Chang, C. (2011). "Feeling ambivalent about going green – Implication For Green Advertising Processing". *Journal of Advertising.* Winter 2011.Vol. 40, Iss 4 pp 19-31.
- Chang, N.J and Fong, C.M (2010). "Green product quality, green corporate image, green customer satisfaction, and green customer loyalty". *African Journal of Business Management.* October 2010.Vol.4 (13), pp.2836-2844.
- Chen, T. B. and Chai,L. T (2010), Attitude towards the environment and green products: consumer perspective, *management science and engineering* vol.4, No 2, pp. 27-39.
- Chitra, K. (April-September 2007). In search of the Green Consumers: A perceptual Study. *Journal of Services Research.* Volume 7, Number 1 pp. 173-191.
- Cone communications "Consumers still purchasing, but may not be "buying" companies' environmental claims". *Trend Tracker* (2012) pp.1-7.
- Churchill, G & Suprenant; 'An investigation into determinants of customer satisfaction', Journal of Awareness and acceptance marketing Research, vol. 19, pp. 491-504, 1992.
- Comrey, AL & Lee; A first course in factor analysis, 2 ed. nd L Erlbaum Associates, New Jersey.

- Datta, S. K., and Ishaswini (2011) Pro-environmenatal Concern Influencing Green Buying: A Study on Indian Consumers, *International Journal of Business and management* Vol.6 No.6, pp. 124-133.
- Deli-Gray, Z., Gillpatrick, T., Marusic, M., Pantelic, D. and Kuruvilla, S.J (October 2010 – March 2011). "Hedonic and Functional Shopping Values and Everyday Product Purchase: Findings from the Indian Study". *International Journal of Business Insights & Tranformation*. Vol. 4, Issue 1, pp. 65-70.
- Elmore, PE & Beggs; Salience of concepts and commitments to extreme judgments in response patterns to teachers, Education, vol. 95, no.4, pp. 325-334.
- Finisterra do Paço, A.M, Lino Barata Raposo, M. & Leal Filho, W. (2009). "Identify the
- green consumer: a segmentation study". *Journal of Targeting, Measurement and Analysis for Awareness and acceptance marketing.* 17, pp. 17-25.
- Florenthal, B. and Arling, P. A (2011). "Do green lifestyle consumers appreciate low involvement green products?". *Awareness and acceptance marketing Management Journal*, Vol.21, Issue 2. pp35-45.
- Gan C., Wee H.Y., Ozanne L.& Kao T. (2008) "Consumer's purchasing behavior towards green products in New Zealand". *Innovative Awareness and acceptance marketing,* Vol. 4, issue 1 pp. 93-102.
- Garland; 'The mid-point on rating scale: is it desirable?' Awareness and acceptance marketing Bulletin, vol. 2, May, pp. 66-70.

- Ghosh, M. (2010) "Green Awareness and acceptance marketing – A changing concept in changing time." *BVIMR Management Edge*, Vol.4, no. 1 pp. 82-92.
- Ginsberg, J. M and Bloom P.N.(2004), Choosing the Right Green Awareness and acceptance marketing Strategy
- Hayes, Bob; Measuring customer satisfaction: survey design, use and statistical analysis methods, ASQ Quality Press, Milwaukee.
- Iacobucci, D, Ostrom, A & Grayson; 'Distinguishing service quality and customer satisfaction: The voice of the consumer', Journal of Consumer Psychology, vol. 4, no. 3, pp. 277-303.
- Kinnear, TC, Taylor, JR, Johnson & Armstrong;Australian awareness and acceptance marketing research,McGraw-Hill, Sydney.
- Hair, JF, Anderson, RE, Tatham, RL & Black, WC; Multivariate data analysis with readings, 4th edn. Prentice-Hall International, Englewood Cliffs, pp.274, 2000
- Massachusetts Institute of Technology (MIT), *Sloan management Review* pp. 79-84
- Hartmann, P. & Apaolaza Ibáñez, V. (2006) "Green Value Added". *Awareness and acceptance marketing Intelligence and Planning.* Vol 24 Iss:7 pp. 673-680.
- Hartmann, P. and Apaolaza-Ibanez, V. (2009). "Green Advertising revisited". *International Journal of Advertising.* Vol .28 No 4, pp.715-739.
- Kumar, P. D. (December 2010) "Green Awareness and acceptance marketing: A Start to Environmental Safety." *Advances in Management*, Vol. 4, no. 12 pp. 59-61.

- Leonidos, L.C., Leonidous, C.N. and Kvasova O (2010), Antecedents and outcomes of consumer environmentally friendly attitudes and behaviour, *Journal of Awareness and acceptance marketingManagement*, Vol. 26 Nos. 13-14, 1319-1344.
- Malhotra, NK; Awareness and acceptance marketing research: An applied orientation, 3rd edn, Prentice Hall, New Jersey.
- Marly, B. R., Levy, M. and Martinex J. (2011). The public Health Implications of consumers' Environmental Concern and Their Willingness to pay for an Eco-Friendly product. *Journal of Consumer Affairs*. Vol.45, No2, pp. 329-343.
- Parasuraman, A, Zeithaml, VA & Berry; 'A conceptual model of service quality and its implications for future research', Journal of Awareness and acceptance marketing, Fall, pp. 41-50
- Picket-Baker, J. and Ozaki R. (2008). "Pro-environmental products: Awareness and acceptance marketing influence on consumer purchase decision". Journal of Consumer Awareness and acceptance marketing, Vol. 25 Iss: 5, pp.281-293.
- Pirani, E. and Secondi, L. (2011). "Eco-Friendly Attitudes: What European Citizens Say and What They Do". *Int. Journal of Environ. Res.*, N0 5, ISSN 1735-6865, pp.67-84.
- Polonsky, M. J. (November 1994). « An Introduction to Green Awareness and acceptance marketing. » *Electronic Green Journal* 1, no. 2, pp.44-53.

- Princen, T. (2008). "Notes on the Theorizing of Global Environmental Politics", *Global Environmental Politics* Vol.8 no1 pp.1-5.
- Rahbar E. and Wahid N. A., (2011) "Investigation of green awareness and acceptance marketing tools' effect on consumers' purchase behavior". *Business Strategy Series*, Vol. 12 Iss: 2, pp.73 – 83.
- Schuhwerk, M.E., and Lefkoff-Hagius, R. (1995). "Green or Non-Green? Does Type of Appeal Matter when Advertising a Green Product?". Journal of Advertising Vol. XXIV, No 2. p. 45-54.
- Sekaran; Research method for business: A skill building approach, John Wiley and Sons, Inc. Selnes,1993, 'An examination of the effect of product performance on brand reputation, satisfaction and loyalty', European Journal of Awareness and acceptance marketing, vol. 27, no. 9, pp. 19–35.
- Tabachnick, BG & Fidell; Using multivariate statistics, 4th edn, Allyn & Bacon, Boston.
- Thøgersen, J. (2011) "Green Shopping: For Selfish Reasons or the Common Good?".*American Behavioral Scientist.* 55 (8) pp.1052-1076.
- Unknown authors. (2009). "Europeans'attitudes towards the issue of sustainable consumption and production". *Flash Eurobarometer Series* no. 256. pp 1-86.
- Van Waterschoot, W. & Van den Bulte, C. (October 1992). The 4P Classification of the Awareness and acceptance marketing Mix Revisited. *Journal of Awareness and acceptance marketing* Vol. 56. pp. 83-93.
- Vernekar, S.S, and Wadhwa, P. (2011). Green Consumption An Empirical Study of Consumers

Attitudes and Perception regarding Eco-Friendly FMCG Products, with special reference to Delhi and NCR Region. *Opinion.* Vol 1, N0 1, December 2011. pp.64-74.

- Wong, V, Turner W. and Stonement (1996), Awareness and acceptance marketing Strategies and Awareness and acceptance marketing Prospects for Environmentally-Friendly Consumers Products, *British Journal of Management*, Vol.7, pp. 263-281.
- Luck, DJ & Rubin; Awareness and acceptance marketing research, 7th edn, Prentice-Hall international, New Jersey.
- Luck, Edwina, M. & Ginanti, A. (2009). "Mapping Consumer's attitudes for future
- sustainable". *Awareness and acceptance marketing Australian and New Zealand Awareness and acceptance marketing Academic.* AANZMAC 2009. pp. 1-8.
- Pallant; SPSS survival manual: a step by step to data analysis using SPSS, Allen & Unwin, Australia.
- Perry, C, Reige, A & Brown; Realism's role among scientific paradigms in awareness and acceptance marketing research, 1998.
- Wannimayake, W.M.C.B. and Randiwela, P. (2008) "Consumer attractiveness towards Green Products of FMCG sector: An empirical study" Oxford Business and Economics Conference Program pp.1-19 june 22-24.
- Wong, YH & Chan; 'Relationship awareness and acceptance marketing in China: Guanxi, favoritism and adaptation', Journal of Business Ethics, vol. 22, pp.107-118.

- Yazdannifard R. and Mercy, I. E (2011). "The Impact of Green Awareness and acceptance marketing on Customer satisfaction and Environmental safety". *International Conference on Computer Communication and Management, Vol.5 pp.637-641.*
- Zikmund; Exploring awareness and acceptance marketing research, 7th edn, Dryden Press, Forth Worth.

Annexure I

QUESTIONNAIRE

1. Do you purchase Environmental friendly Products?
   - ☐ Yes (pls. respond question no 3 onwards)
   - ☐ No (pls. respond question no 2)

2. Reasons for not purchasing Environmental friendly Products.
   - ☐ They are expensive
   - ☐ Claim is not trustworthy
   - ☐ Inferior Quality as compare to chemical counterpart
   - ☐ Less variety
   - ☐ Non standardization of products
   - ☐ Less shelf-life
   - ☐ Any other _____

3. Do you get motivated to buy a product focusing on Organic/ Environmental friendly essence in them?
   - ☐ Yes
   - ☐ No

4. How frequently you purchase Environmental friendly Products?
   - ☐ Always
   - ☐ Most of the time
   - ☐ Sometime
   - ☐ Rarely
   - ☐ Never

5. Which particular store you like to purchase Environmental friendly Textile Products from?
   - ☐ Khadi Gram Udyog
   - ☐ Fab India
   - ☐ Earth
   - ☐ Anokhi
   - ☐ Cottage Emporium
   - ☐ State Emporiums
   - ☐ Siyaram's
   - ☐ Fibre2Fashion
   - ☐ Good Earth
   - ☐ Grassroots by Anita Dongre
   - ☐ Any other _____

6. Which particular store you like to purchase Environmental friendly FMCG Products from?
   - ☐ Khadi Gram Udyog
   - ☐ Fab India
   - ☐ Body Shop
   - ☐ Patanjali
   - ☐ Cottage Emporium
   - ☐ State Emporiums
   - ☐ Organic India
   - ☐ Natures
   - ☐ Himalaya

- ☐ Eco India
- ☐ Good Earth
- ☐ Any other _____

7. According to you, Environmental friendly Products are the ones which:
   - ☐ Produced by natural resources
   - ☐ Help in conserving natural resources
   - ☐ Complete production chain is controlled from an environmental perspective.
   - ☐ Are manufactured using green technology that caused no environmental hazards
   - ☐ Are recyclable, reusable and bio-degradable
   - ☐ Have herbal and natural ingredients
   - ☐ Are non-toxic, minimize carbon emission and contain recyclable ingredients
   - ☐ Do not harm or pollute the environment
   - ☐ Are not tested on animals
   - ☐ Have eco-friendly packaging
   - ☐ Any other _____

8. How often do you purchase following Environmental friendly Products?

S.N	Products	Never	Rarely	Some time	Most of the time	Always
1	Tea/ Coffee					
2	Juice					
3	Soft-Drink					
4	Bottled water					
5	Health Drinks					
6	Confectionary					

7	Staples / Cereals/ Spices					
8	Bakery (Bread/ Biscuit/Cakes)					
9	Snack food					
10	Chocolates					
11	Ice-creams					
12	Processed fruits					
13	Vegetables					
14	Meat / Dairy product					
15	Dental care					
16	Skin care / Hair care					
17	Paints & Polishes					
18	Personal Wash					
19	Cosmetics					
20	Toiletries					
21	Talcums					
22	Deodorants / Perfumes					
23	Washing/ Laundry Soaps detergents					
24	Kitchen utensil cleaners					
25	Floor cleaners					
27	Air-freshener					
28	Mosquito repellents					
29	Suite/Kurta					
30	Saree					
31	Lowers					
32	Bedsheets/ Beddings					

33	Shawls					
34	Curtains					
35	Towels					
36	Upholstery					
37	Kitchen Cloths					
38	Napkins					
39	Floor Mats					
40	Scarf					
41	Dining Mats/ Napkins					
42	Table Cloths					

To what extent you agree with the following statement. (1 –strongly disagree and 5 –strongly agree)

1. I think that the green product does not carry harmful ingredients like insecticide/ pesticide/ plastic/non-biodegradable or non-recyclable input materials.
2. I think green process of production respects the environmental health
3. I think that environment friendly products may cause quite less harm our environment as compared to its chemically laded counterpart.
4. I believe that with advent of latest technologies, use of eco-friendly products will increase and improve the efficiency in reduce-recycle reuse chain to support greener environment.
5. The seal of quality organizations will further support & increase purchase of Eco- for greener environment.
6. I think that it is ethical on my part if I support the green products because they have (i) well defined origin (ii) are natural & safe foe consumption and in disposal.
7. There is respect for our nature in doing so.
8. My ethics, Values, norms and culture motivates me to stretch a little to be ethical toward the environment I live in.
9. I am interested purchasing eco whose by product are used for renewable initiative.
10. The purchase of green product is for the reason that I want to do may bit for our environment.
11. I wish to purchase environment friendly nature product that do not harm the environment and society.
12. I can purchase the green product if the packaging & labeling information provided is sufficing my clearly doubts.
13. I am in favor of the green product that any stages have used the sustainability process from renewable success.

14. Quality of green product is Good.
15. Natural ingredients as claimed.
16. Keeps promise of being safe for conception.
17. Does not harm health.
18. Perceived as good value for money preposition even if is priced little higher.
19. Advantage outweigh the disadvantage so can pay more for sake of owes and society
20. I take conscious decision of buying the greener attentive product for the sake of my duty towards the environment.
21. I help others by persuading them to be environmentally cautious citizens.
22. I believe that my simple/Small action can cause big impact towards the eco-system balance.
23. I am inspired by the other who works toward having a automatable society.
24. I believe in the individual empowerment necessary for it.
25. In unrelieved to buy environmentally safe products for my personal happiness and satisfaction.
26. It improves my quality of wellbeing in life.
27. It gives a trek message to my peer group.
28. It symbolizes the culture and values I possess.
29. It shone my deep rootedness to cultural/historical roots.
30. It gives me more acceptances in my primary/Secondary /treasury groups.
31. I feel pride in explaining eh pros of eco-product, its information and how I am doing my contribution to society at larger
32. I respect the wish of my other us my social group.
33. I support the decisions taken during purchase by any peer-group/Family
34. The problem of depleting natural resources can be resolved by my message to society for green.

35. I am worried about the environment and what holds in the future.
36. I think my initiative can lessen the severely upset environmental balance.
37. Humans must live in harmony with the eco-systems/ nature in order to survive.
38. I make special effort to reduce the use of paper non-recyclable product to pass on this message to the society.
39. I don't mind spending on eco/green product if it convinces me on consistency and trustworthiness of brand being herbal.
40. I can represent my socio-economic status before my poual groups by purchasing green.
41. I am competent to pay more
42. It is my obligation to pay more for environment
43. My culture/ Value makes me pay more for environment
44. I am status conscious
45. I can catalyze sustainable environment by paying and enchase to the engagement & involvement & Encouragement in working as a catalyst to enable green product advantage ones non-green and hence witting to pay because of its various superior outcomes.
46. Trustworthiness of marketer's proofs of being eco-friendly.
47. Price as major factor in slow growth of sales of green products.
48. Transparency of organization using green awareness and acceptance marketing concept since long.
49. Employees/Customer feel that routine can be affected by implementing green concept.
50. It is difficult for all companies to adapt to green initiatives compliantly.
51. Transparency n contents of the product labeled green.

Respondent Profile

Gender	☐ Male	☐ Female	
Age Group	☐ <18	☐ 18-25	☐ 25-35
	☐ 35-45	☐ >50	
Personal Income(Annual)	☐ Dependent	☐ <5Lacs	☐ 5-10Lacs
	☐ 10-15Lacs	☐ >15Lacs	
Family Income(Annual)	☐ 5-10Lacs	☐ 10-15Lacs	☐ 15-25Lacs
	☐ >25Lacs		
Occupation	☐ Student	☐ Homemaker	☐ Service
	☐ Business		

Curriculum Vitae

DR. APARNA GOYAL

SUMMARY

Dr. Aparna Goyal has been associated with Amity University, Noida since 2005 in the area of Marketing, Retailing, Strategy, Branding and Advertising as Associate Professor in FMS. Known for her energy, enthusiasm, positivity, passion and empathy, she headed various main-stream strategic tasks, remained Program Coordinator for PGDM, Chairperson of MBA Program, Chairperson of Admissions, Corporate Interaction Cell and Core coordinator for the Admissions at Amity Business School, Amity University, the renowned education group of India, accredited by reputed agencies and placed amongst world's best B-Schools. She began her career in Product & Brand Management Marketing at Alps Industries Ltd., before joining academics. Dr. Goyal is a Bachelor of Science, Education Graduate, Masters' in Organic Chemistry, and Diploma in Computers. She was awarded Ph.D (in Green Marketing) from Sai Nath University in

2012. She has submitted her second Ph.D Thesis (Eco-Sustainable Sensory Marketing) at Amity University in 2016. Teaching and research domains are Marketing, Branding, Retailing, Consumer Behaviour, Advertising, Green Marketing & CSR, and Digital Online Marketing. She has 72 publications in high Indexing, peer-reviewed, refereed, high impact factor, ISSN, DOI – Journals like Scopus, Index Copernicus, Thomson Reuters Researcher ID, Google Scholar, UGC approved, PubMed, Indian Citation Index and the like - to her credit, including books, international research publications, proceedings, seminars, presentations, book chapters, book reviews and case studies. She presented research papers in national & international conferences in Singapore, Kolkata, Chandigarh, Varanasi, Indonesia and Delhi NCR. She designed and delivered numerous courses and program SOP's for general and executive management programmes as a facilitator and Program Coordinator for In-company and conducted workshops for Higher Education implementation and evaluation.

EDUCATION

- Ph.D., Amity University (Submitted in 2016), Major Field: Marketing and HR
- Ph.D., Sai Nath University (Awarded on 23-12-2012), Major Field: Marketing
- P.G.D.M., Institute of Technology and Science, Area of Concentration: Marketing and HR -1998 (74%)
- M.Sc. (Organic Chemistry), CCS University – 1996 (69%)

- B.Ed. (Bachelors in Education), CCS University – 2000 (70%)
- B.Sc. (Bachelors in Science), CCS University – 1994 (66%)
- Diploma in Software Management, Aptech Computers- 1993 (81%)
- Senior Secondary (CBSE), Delhi Public School Ghaziabad (1991) – 76%
- Higher Secondary (CBSE), Delhi Public School Ghaziabad (1989) – 78%

PROFESSIONAL EXPERIENCE

- 2005- till date- Associate Professor, Amity University Noida
- 2001-2005- Lecturer, Institute of Technology and Science Ghaziabad
- 1998-2001- Marketing Executive, Alps Industries Ltd. Sahibabad
- 1996-1998- VMLG PG College, Ghaziabad – Visiting Faculty

SCHOLASTIC ACHIEVEMENTS

- All India Management Association (Honorable Member since 2008)
- Honorable Reviewer of SCOPUS ELSEVIER Journal – Ecological Indicators
- Honorable Member of the Editorial Board and Reviewer of the International Journal of Research in Engineering & Innovation (International peer-reviewed refereed multi-disciplinary, Indexed in

Google Scholar, Academia.edu, Thomson Reuters Research ID, Scribd, Slide share, Bibsonomy and more)
- Finalist, Best Paper Presenter Award in International Conference ICFBE 2017, President University, Indonesia
- Runner-up, National Conference on Roots of Indian Management at Rajarshi Institute, Varanasi
- Honorable Guest for Melting-pot Forum for Industry-Academia Interaction
- Reviewer for World Journal of Engineering Research & Technology for one year
- 2016 RI Publications Foundation published research paper, SCOPUS indexed that have made lasting contributions to the discipline of management
- Thesis on Green Marketing accepted for publication with MoU signed with prestigious publication house of U.S.A., Partridge Publications
- Lifetime Membership of AIAER (All India Association for Educational Research)
- Imparted Training to Organizations like Tata Motors, Central Warehousing Corporation
- Conducted Market Research for Hunter Douglas, U&I Exports, Ricky International, Johnson & Johnson, Alps Industries Ltd.
- Semester Abroad Programme at Amity University Singapore
- Imparted trainings at Central Warehousing Corporation CWC, NTPC, Le Pashmina, Hunter Douglas, U&I Exports, Kanika's, Nokia, Alps Industries at the Middle Management Level.

SPECIAL ACHIEVEMENTS

- **Program Leader (MBA Class of 2005, 2009 & 2016)** – For each batch, PL is the academic parent of 400+ students of MBA batches. Major roles included, supervising and coordinating the administration and governance of studies, serving as a nodal point of contact for all students and acting as liaison among students, program faculty, and the college and the University. As PL, headed the college team w.r.t. education-related policies, deadlines, and programs as appropriate; and forwarding recommendations, nominations, and other information from the faculty to the appropriate collegiate and FMS administrations. Implemented discipline criteria upon entry to exit with acceptable progress rules through the program and the grounds for student's termination. Arranged for the review of, and monitored the progress of student applications and petitions; oriented and counselled students with respect to program and degree requirements; enforced regulations of the University. Important responsibility included monitoring the maintenance of graduate student records and the annual student evaluation process, providing periodic reports on the program and data to the collegiate dean.
- **Admissions Coordinator** – She has performed a number of management, outreach, and promotion roles. Dr. Goyal knows university policies and procedures, as well as understands the long-term goals and curricula of their departments, and uses this knowledge to guarantee these goals met, while

she works to recruit students to MBA program. Dr. Goyal recruited and retained students into the MBA program including program marketing, developed and marketed MBA programs for students, corporate, government and professional audiences. She interacts with faculty and other business leaders to continually enhance and improve the quality of the students selected by way of ensuring curriculum and the programs being competitively distinct.

- **Teaching- Learning Pedagogy** – Executing the concept of learning by doing. Active application and use of learning environments allow students to discuss, talk and listen; read, write, and reflect as they approach course content through problem-solving exercises, informal small groups, simulations, case studies, role playing, mental imagery, critical thinking, implicit & integrated learning, empathetic learning, peer- tutoring, thematic teaching learning, meta-cognitive awareness, prediction & inference based applications, demonstration learning, selective attention, cognitive processing, concept mapping group activities, gallery walks, concept sketch, jig-saw, quizzes, diagnostician coaching, market visit projects, reflective action, holistic and experiential learning, other activities students to apply what they are learning
- **Objectives** of a particular evaluation process and the mechanisms used in a way that PLOs and SLOs objectives reached. Evaluation process consists of improvement and accountability. Ensured, that there is a close alignment between the learning goals, the teaching and learning activities aimed

at meeting learning goals and the assessment tasks used to assess whether learning goals have been met. Current best practice are used includes assessment which is aligned to learning goals which focus not only on content knowledge but also on process and capabilities.

- **Demonstration of subject-area knowledge**, pedagogical knowledge, and professional teaching ability is the sacrosanct goal. Current efforts promise in ensuring that we enter the class with all the necessary intellectual input, and choose a high standard of preparation route to the classroom. Each PLO & SLO of OAP is aligned with the criteria used for class delivery as per the topic in the module.
- **RESEARCH GUIDANCE FOR DISSERTATION/ INTERNSHIP/ LIVE PROJECTS – 256** (TILL NOV'2017)
- **In-charge - Student progress** regularly and have contingency plans if something goes wrong. Her outcome plans replicate real world activity. Students have guidelines on what is acceptable and what is not acceptable in terms of their collaborative versus individual work and assessment. Class is alerted as to how incidents of academic misconduct, such as plagiarism, to be avoided. Acts congruently with, and advocates for, the ethical standards of chosen discipline. It is refinement of technique wherever essential. The core purpose Dr. Goyal keeps in mind, is to strengthen the knowledge, skills, dispositions, and classroom practices through continuous engagement in learning, skill acquisition, and refinements to practice is

essential for meeting student learning needs. Outcomes assessed as per standards laid down by our University and help us in determining whether students' are acquiring and applying the content, skills, and dispositions necessary to meet standards for student learning. We identify any additional learning that a student manager may need. In-class culture and climate considered to ensure that the learning outcome environment is supportive of ongoing professional learning. Collaboration among peer-groups and teacher reflective practice as characteristics of supportive learning outcome environment

- **OAP based on comprehensive standards of practice**, which could include classroom observations, evaluations, and teacher-student conferences, portfolios, evidence binders, conference presentations, and instructional additional inputs from current economy to show demonstrated attainment and use of new knowledge and skills. Use of information provided through formative assessments, peer reviews, professional learning communities, and other forms of feedback and support is utilized which becomes an evidence of student growth and learning based on multiple measures as defined by the University. Professional opportunities had been, through memberships, given to students. Under Governance, motivation to learn is more likely when we have provided course clarity, *task clarity,* and clear learning goal and know how. Learning goals and assessments are meaningful and worth learning and also provides the potential for success.

- **Interpersonal competencies** enhanced; include communication skills, cooperative learning, and courtesy. Intrapersonal competencies include work habits such as organization, time management, and persistence is given due weightage. I encourage rewards as effective teaching practices based on student learning outcomes because it enables students to identify their own strengths and weaknesses and determine the kinds of information they need to correct their learning deficiencies and misconceptions. This evaluation properly employed; students learn that they can engage in self-assessment and continuous improvement of performance throughout their lives. Fostering academic honesty is being emphasized by me and been stated in clear, easy to understand terms. Dr. Goyal provides opportunities to demonstrate progress—more than one exam grade. I also let students develop class norms that support honesty. Many students would rather not cheat but feel threatened if they think that others are succeeding at cheating and are getting better grades.
- **Mentoring, monitoring incentives, oral or written communication** of praise/ achievement and academic target setting that in turn helps in identifying strengths and weaknesses for further professional development. The aim is to perform best to enhance student learning, includes developing a culture of mutually providing and receiving feedback, clear individual and collective objectives with regard to improving teaching, planning and preparation. I believe in demonstrating knowledge of students, selecting instructional goals, designing

coherent instruction, assessing student learning and facilitating an environment of respect and rapport, establishing a culture for learning, managing classroom procedures, managing student behaviour.

- **Reflecting on teaching learning strategies** include, maintaining accurate records while contributing to institute and University with professionalism. Dr. Goyal feel sensitized, take the ownership of student assessment, and accept it as an integral part of teaching and learning. I offer regular descriptive feedback on my student progress. Classroom assessment practice informs teacher lesson planning, assists with students learning needs, and engages students in the assessment process. Dr. Goyal deals with Academic Dishonesty by employing a proactive approach to limit incidents of academic dishonesty. I strategize on how to prevent it and what to do about it when it happens. I provide instructional and assessment practices to conferences / seminars, which they should attend, to keep track of their progress towards being placeable and provide them with steps towards being engaged in corporate communication for ultimate success through PR. I maintain separate diary per semester for my remarks on class/ student problem or behaviour, etc. so that I remember my observations and provide fair achievement record for grading.
- **Coordinator Alumni Meet** – Communicates with alumni and professionals, and collaborate with other departments in your university. She also forms partnerships with community businesses, working with them to fund your program initiatives.

education-related policies, deadlines, and programs as appropriate; and forwarding recommendations, nominations, and other information from the faculty to the appropriate collegiate and University administrations. Working tirelessly towards building relationship with community and business leaders in the region; directs the preparation of directives to division or department supervisors outlining program, or operational changes to be implemented. Participates in various committees, workshops and trainings; coordinates, schedules and conducts meetings.

- **Area Coordinator (Marketing Domain)** –An active frontier in MBA Expansion Committee; Curriculum Development Committee; Board of Studies; Course Advisory Committee for Strategy Review and Academic Performance Committee. She has been responsible for supervising and directing the workflow of marketing, advertising promotions by assigning job tasks, facilitating interdepartmental communications and managing external communications. Her expertise allows FMS to allocate institutional resources to maximize collaboration, efficiency and creativity in the building and maintenance of consistent branding or corporate identity across marketing and public relations channels. She is responsible for staying up to date on industry trends to hand hold the team members. The marketing management side of the job involved the development and refinement of marketing goals, pricing strategies, promotional activities and branding in consultation with marketing staff and placement center of the

college. Directing market analysis and research to identify trends and opportunities is another job duty she had been performing, with the ultimate goal of enhancing the placement opportunities.
- **Placement Counsellor & Advisor for Amity Global Business School** – Initiatives -
  - Mentoring
  - Student's Counselling
  - Grievance Handling
  - Coaching
  - Initiated Corporate & Industry Counselling Center
  - Placements In-charge
  - Conducted Mock GD & PI
  - Organized Conferences and Seminars
  - Anti-ragging & Disciplinary Committee
  - Human values Quarter
  - Amity Youth Festivals
  - Sangathan- Annual Sports & Cultural Festival
  - Corporate Meets & Lunches
  - Sponsorships
  - Esteemed Judge at various Inter & Intra Institute level competitions
  - Co-chaired Academic Strategic SOPs
  - Head – Marketing & Retail Club
  - Member of Proctorial Board
  - Head of Personality Grooming Sessions
- **Coordinator – Accreditation Team –** As part of main accrediting team at Amity University, various recognition/approval attained by performing with complete & dedicated involvement in recognition by the UGC, AIU and member of the Association of Commonwealth Universities. Her team attained

the coveted QS University Rankings and BRICS 2015, where in Amity ranked amongst top 94 Indian Universities and 400 Universities of BRICS countries. As coordinator, her contribution to accreditation by the NAAC (National Accreditation and Assessment Council) with grade A in 2012 had been commendable. Not only this, number of management programs were accredited by ACBSP (Accreditation Council for Business Schools and Programs); accredited by the Western Association of Schools and Colleges (WASC). Ranked 46 in India by the National Institutional Ranking Framework engineering ranking for 2017 and 65 in India by Outlook India's "Top 100 Colleges In 2017". National Assessment and Accreditation Council (NAAC) with "A" Grade recognized Amity due to mammoth work done, including the same commitment for Scientific and Research Organization by DISR, Ministry of Science Technology, Government of India accreditation as well. Accreditation by ASIC, UK (Accreditation Service for International Colleges) with rare distinction as "Premier College", is another feather to her cap.

- **Research Papers / Articles / PROCEEDINGS/ CASES/ BOOK CHAPTER/ REVIEW/ Conferences / Seminars / FDP – 72** (Please See Annexure – I Enclosed)

Personal Details –

Name   – Aparna P. Goyal
DOB   – August 14, 1974
Add   – 220, Navneete Aptt., IInd Floor, Plot No. 51, Ipextn, Patparganj, Delhi 92
Email   – meaparna148@gmai.com; 148meap@gmail.com
Mo   – 7065800350; 8130620375

Signatures & Date

Annexure II

PUBLISHED PAPERS

- Aparna Goyal, "An Exploratory Study for Consumer Buying Behavior for Environment-Friendly Products in Fmcg Sector" Indian Journal of Applied Research 2249-555X [2014] Vol4, Issue 4 page01-03 © ICI, GENAMICS, DRJI, ULRICH'S PERIODICALS DIRECTORY, World Cat, iifs, Index Copernicus International, EZB, I2OR, INNO SPACE, INDIAN Science, DJOF, Research Gate, EyeSpource International, Open J- Gate, ISI, COSMOS, Quality Factor
- Dr. Aparna P. Goyal, Dr. Teena Bagga, Dr.Sanjeev Bansal, "Impact of Increasing Trend of Online Marketing on Consumer Buying behaviour: FMCG Brands in Indian Scenario" International Journal of Engineering Technology, Management and Applied Sciences 2349-4476 [2016] Volume 04 - Issue 05 page218-229 © DIIF, Google Scholar, CiteSeer,Q. Sensei"

- Aparna Goyal, "Consumer Purchase Attitude And Behaviour With Respect to Brand Preference and Brand Loyalty For Environmentally Favourable Products" Paripex-Indian Journal of Research 2250-1991 [2014] Volume 3: Issue 5 page135-136 © Google SCholar, ULRICH'S PERIODICALS DIRECTORY, JOUR Informatice, ijIndex, INNO SPACE, Eye Source, ISI, World Cat, Index Copernicus, Open J Gate, DJOF, Cite Factor, GENAMICS, EZ3 Electronic Journal Library, ICI, Indian Science, Socol@r,DRjI,I2OR,COSMOS, CSA Illustra
- Dr. Aparna Goyal, "The Buying Habits of Fast Moving Consumer Goods: A Case Study on the Slum Area in Delhi" International Journal of Scientific Research 2277 - 8179 [2014] Volume 3: Issue 7 page258-259 © ULRICH'S PERIODICALS DIRECTORY, Research Gate, Eye Source, ISI, World Cat, Index Copernicus, Open J Gate, DJOF, Cite Factor, GENAMICS, EZ3 Electronic Journal Library, ICI, Indian Science, Socol@r, PUBMed,
- Dr. Aparna Goyal, "Eco-friendly Marketing: The market potential for sustainably managed Wooden Products - Home & Office Furniture, in Indian Scenario" Global Journal for Research Analysis 2277-8160 [2014] Volume 3, Issue 3 page116-119 © Google SCholar, ULRICH'S PERIODICALS DIRECTORY, JOUR Informatice, ijIndex, INNO SPACE, Eye Source, ISI, World Cat, Index Copernicus, Open J Gate, DJOF, Cite Factor, GENAMICS, EZ3 Electronic Journal Library, ICI, Indian Science, Socol@r,DRjI,I2OR,COSMOS, CSA Illustra

- Dr. Aparna P. Goyal, "Ethical Eco-Fashion-Future or Utopia: Role of Sustainability Marketing" International Journal of Engineering Sciences & Management Research 2349-6193 [2016] Vol.3, Issue 12 page117-128 © Thomson Reuter Researcher-ID, Google Scholar, doi, Cite Seer, iisi, Journal Index.net, St George's University of London, DRJI, Academia.edu.
- Aparna Goyal, "Intentions and Perception of Customers with regard to Ecologically Sustainable Eco-Green Initiatives" International Journal of Engineering Science Technology And Research (IJESTR) 2456-0464 [2017] Volume 2 Issue 1 page05 - 18 © SCOPUS, EBSCO, BioMED, DOSI, ROAD, Journal Index, OAJI, Google Schlolar, DOAJ, Cite factor, Scholaster, Scopus
- Aparna Goyal, "A Study of Consumer Perceptions and Purchase Behaviour Trends Towards Digital online Buying Behaviour of customers from Different age-groups" International Education and Research Journal 2454-9916 [2017] Vol.3, No 1 page95-100 © ICI, Google Scholar, PUBMED, CiteFactor, OCLC worldcat,ISI, ISSUU, I2OR, Internet Archive, ULRICH's, Academia.edu, CiteUlike, IJSRAF
- "Teena Bagga, Aparna Goyal, Sanjeev Bansal, "An Investigative Study of the Mobile Operating System and Handset Preference" Indian Journal of Science and Technology 0974-6846 ISSN (O): 0974-5645 [2016] Vol.9, Issue 35 page1-14 © Google Scholar, Ulrich Web, Cabell's Directory, SCOPUS,Open J Gate, Jour Informatics, MicrosoftThomson Reuters "Web of Science"

- Aparna P. Goyal, "Personalized Niche Online Marketing to Generate and Sustain E-Loyalty Among the Millennial Generation" International Journal of Current Advanced Research 2319-6505 [2017] Volume 6; Issue 3 page2421-2428 © UGC Journal NO:43892, Thomson Reuter Researcher-ID, Pubmed, THOMSON REUTERS RNDNOTE, ICMJE
- Dr. Aparna Goyal, "Study of Consumer Purchase Behaviour in the Context of Organized Retail Outlets" International Journal of Recent Research in Commerce Economics and Management 2349-7807 [2017] Vol. 4, Issue 1 page9-18 © YUDUfree,PDFCAST.org,author stream,Google, Slideshare
- Aparna Goyal, "Study of Consumer Purchase Behaviour in the context of Organized Retail Outlets of Reliance Fresh in Delhi" International Journal of Current Innovation Research 2395 – 5775 [2017] Vol. 3, Issue 01 page554-561 © Site Factor, Indian Science, Science Central.com,DRJI, Google Scholoar, Index Copernicus,Myinfoline,SiS, GIF, Research Gate
- Dr. Aparna P. Goyal, "Study of Consumer Purchase Behaviour in the Context of Organized Retail Outlets of Pantaloons in Delhi" International Journal of EngineeringSciences & Management Research 2349-6193 [2017] Vol. 4(1) page27-38 © Thomson Reuter Researcher-ID, Google Scholar, doi, Cite Seer, iisi, Journal Index.net, St George's University of London, DRJI, Academia.edu.
- Aparna Prashant Goyal, "To study use of Sensory-Neuro Marketing Strategy on Consumer Buying

Behaviour Towards Eco-friendly Organic Green Products" World Journal of Engineering Research and Technology 2454-695X [2017] Vol. 3, Issue 3 page254 -286 © Google Scholar, THOMSON REUTERS Researcher ID: M-6054-2017, UGC APPROVED JOURNAL LIST [UGC JOURNAL NO: 47231], Chemical Abstract Services Databases (CAS databases).

- CNKI Scholar, Directory of Open Access Journal (DOAJ), Sweden Index Copernicus International Ltd, Poland, Cite Factor,Scribd, Crossref., Science Library Index, scientific World Index, ROAD Directory of Open Access Scholarly Resources, COPAC Root Indexing,MIAR 2014 LIVE,Information Matrix for the Analysis of Journals, ResearchGate,ICI JOURNALS MASTER LIST, Mendeley, National Library of Medicine, MEDLINE/PubMed Data Element, Pakistan Academic Research (Growing Knowledge for Feature), Electronic journal library, Academic Journals Database., Library.USASK; University of Saskatchewan, International Society for Research Activity, Urlich's Periodicals Directory, Proquest, USA, Open-J-Gate, India, NewJour-Georgetown University Library, USA, The Open Access Digital Library, USA, Lenide Journals, Dayang journal system, Korea, Science central, USA,San Jose State University Library, USA Pharma Trendz International, Pharmpedia, Canada, Chemical Abstracts Service (CAS), USA, SOCOLAR, China, Google Scholar,Indian Science, LexiNexis, Research Bible – Journal Seeker, GFMER,Electronic Journals Library (EZB), ZDB-Database, Germany, Rubriq

Beta, Academia, Refertus, eannu, The Journal database (TJDB) and in TutorGig Western

- Dr. Aparana Prashant Goyal, "TO UNDERSTAND THE CONSUMER PURCHASE PERCEPTIONS AND REASONS FOR MOTIVATION TO BUY THROUGH ONLINE MARKETING STES" International Journal OF Engineering Sciences &Management Research 2349-6193 [2017] Volume 4, Issue 4 page45-50 © Thomson Reuter Researcher-ID, Google Scholar, doi, Cite Seer, iisi, Journal Index.net, St George's University of London, DRJI, Academia.edu.

- Dr. Aparna Goyal, Prof. (Dr.) Sanjeev Bansal, "Study of Organic Agricultural Practices for Improving Environment Sustainability in India" International Journal of Scientific and Research Publications 2250-3153 [2017] Volume 7, Issue 5 page636-648 © Google Scholar, BASE (Bielefeld University Library) and OARD (Open Access Research Database)

- Dr. Aparna Goyal and Dr. Sanjeev Bansal, "Digital Platforms & Natural eco-Digital eco Strategy – Top Brands Igniting it Through Environment Sustainable Revolution" International Education and Research Journal 2454-9916 [2017] Vol. 3, Issue 5 page497-504 © ICI, Google Scholar, PUBMED, CiteFactor, OCLC worldcat,ISI, ISSUU, I2OR, Internet Archive, ULRICH's, Academia.edu, CiteUlike, IJSRAF, UGC approved

- Dr. Aparna Goyal, "TO STUDY AND EXPLORE THE CUSTOMER ATTITUDE TOWARDS SAFER NATURAL PRODUCTS QUALITATIVELY FOR FMCG FOODSECTOR

ECOBRANDINGIN INDIAN SCENARIO" International Journal of Research Science & Management 2349-5197 [2017] Vol 4, Issue 7 page08-27 © Thomson Reuter Researcher-ID, academia. Edu, siteSeer

- Aparna Goyal, "To Explore Customer Purchase Trend in Retail Food Supermarkets w.r.t. Branding Information on the Packs: A Qualitative Research" World Journal of Engineering Research and Technology 2454-695X [2017] Vol-3 Issue 4 page441 -476 © Google Scholar, THOMSON REUTERS Researcher ID: M-6054-2017, UGC APPROVED JOURNAL LIST [UGC JOURNAL NO: 47231], Chemical Abstract Services Databases (CAS databases).

- CNKI Scholar,Directory of Open Access Journal (DOAJ), Sweden Index Copernicus International Ltd, Poland, Cite Factor,Scribd, CrossRef., Science Library Index, scientific World Index, ROAD Directory of Open Access Scholarly Resources, COPAC Root Indexing,MIAR 2014 LIVE,Information Matrix for the Analysis of Journals, ResearchGate,ICI JOURNALS MASTER LIST, Mendeley, National Library of Medicine, MEDLINE/PubMed Data Element, Pakistan Academic Research (Growing Knowledge for Feature), Electronic journal library, Academic Journals Database., Library.USASK; University of Saskatchewan, International Society for Research Activity, Urlich's Periodicals Directory, Proquest, USA, Open-J-Gate, India, NewJour-Georgetown University Library, USA, The Open Access Digital Library, USA, Lenide Journals, Dayang journal

system, Korea, Science central, USA,San Jose State University Library, USA
- Pharma Trendz International, Pharmpedia, Canada, Chemical Abstracts Service (CAS), USA, SOCOLAR, China, Google Scholar,Indian Science, LexiNexis, Research Bible – Journal Seeker, GFMER,Electronic Journals Library (EZB), ZDB-Database, Germany, Rubriq Beta, Academia, Refertus, eannu, The Journal database (TJDB) and in TutorGig Western
- Dr. Aparna Goyal, "Radical Incremental Innovation in Products and Brands: Factors Influencing Customer towards their Adoption" International Journal of Engineering Research and Applications 2248-9622 [2017] Vol. 7, Issue 8, (Part -5) page13-31 © UGC Approved - SN 4525, 80.82 Copernicous Index, Google Scholor
- Dr. Aparna Goyal, "Online trends towards buying behaviour of customers from different segments: Study in NCR" Indian Journal of Applied Research 2249-555X [2017] Vol 07, Issue 04 page38-44 © ICI, GENAMICS, DRJI, ULRICH'S PERIODICALS DIRECTORY, World Cat, iifs, Index Copernicus International, EZB, I2OR, INNO SPACE, INDIAN Science, DJOF, Research Gate, EyeSpource International, Open J- Gate, ISI, COSMOS, Quality Factor
- "Dr. Aparna Goyal, "Marketing Strategy And Plan– Role in Brand Building of Kama Vs. Kara Ayurveda Nose Pores Remover: A Study in NCR" International Refereed Journal of Engineering and Science ISSN (Online) 2319-183X, (Print) 2319-1821 [2017] Volume 6, Issue 8 (August 2017) page70-80

© UGC Approved - SN 4343, opernicous Index, Google Scholor, Open J-Gate, Cabell's Directories, Jour Infromatics, ULRICHSWEB"

- Aparna Goyal, "A Study of Psychological Perspective of Customers w.r.t. Rising Digital Retailing" International Journal of Recent Scientific Research ISSN No: 0976-3031 [2017] Vol 8, Issue 6 page17708-17718 © NCBI, PuB Med, THOMSON REUTERS, ENDNOTE, ICMJRdoi. Cross reference.

- Aparna Goyal, "To Study the Consumer Perceptions towards building Self-motivation through various Online Marketing Strategies" Global Journal of Advanced Engineering Technologies and Sciences 2349-0292 [2017] Vol.4, Issue 4 page60-65 © IISI, CiteFactor,BASE,CHS,IISS. Cold Spring harbor niversity.

- Aparna Prashant Goyal, "To study the Perception of Digital Buyer's reason for hesitation and slow-adaptation of Unified Payment Interfaces" World Journal of Engineering Research and Technology 2454-695X [2017] Vol. 3, Issue 3 page322 -347 © Google Scholar, THOMSON REUTERS Researcher ID: M-6054-2017, UGC APPROVED JOURNAL LIST [UGC JOURNAL NO: 47231], Chemical Abstract Services Databases (CAS databases)

- CNKI Scholar, Directory of Open Access Journal (DOAJ), Sweden Index Copernicus International Ltd, Poland, Cite Factor, Scribd, CrossRef., Science Library Index, scientific World Index, ROAD Directory of Open Access Scholarly Resources, COPAC Root Indexing,MIAR

2014 LIVE,Information Matrix for the Analysis of Journals, ResearchGate,ICI JOURNALS MASTER LIST, Mendeley, National Library of Medicine, MEDLINE/PubMed Data Element, Pakistan Academic Research (Growing Knowledge for Feature), Electronic journal library, Academic Journals Database., Library.USASK; University of Saskatchewan, International Society for Research Activity, Urlich's Periodicals Directory, Proquest, USA, Open-J-Gate, India, NewJour-Georgetown University Library, USA, The Open Access Digital Library, USA, Lenide Journals, Dayang journal system, Korea, Science central, USA,San Jose State University Library, USA Pharma Trendz International, Pharmpedia, Canada, Chemical Abstracts Service (CAS), USA, SOCOLAR, China, Google Scholar,Indian Science, LexiNexis, Research Bible – Journal Seeker, GFMER,Electronic Journals Library (EZB), ZDB-Database, Germany, Rubriq Beta, Academia, Refertus, eannu, The Journal database (TJDB) and in TutorGig Western

- Aparna Goyal, "To study the Operating Procedures of Stock Markets and suggest strategies for transparency in Transactions." International Journal of Current Advanced Research ISSN: O: 2319-6475, ISSN: P: 2319-6506 [2017] Volume 6; Issue 6 page4370-4382 © UGC Journal NO:43892, Thomson Reuter Researcher-ID, PubMed, THOMSON REUTERS RNDNOTE, ICMJE

- Dr. Aparna Prashant Goyal, "Study of Customer Perception towards Services Provided by Five Star Hotels in Indian Context" International Refereed Journal of Engineering and Science ISSN (Online)

2319-183X, (Print) 2319-1821 [2017] Volume 6, Issue 8 (August 2017) page37-46 © Google Scholar,Index Copernicus,Ulrich Web,bell's Directory, Open J Gate Jour Informatics.
- Dr. Aparna Goyal, Dr. Teena Bagga, "Mobile Banking Usage: An empirical research in Delhi/NCR" Indian Journal Of Applied Research 2249-555X [2016] Vol.9 page1-14 © Google Scholar, CNKI Scholar, DOAJ Sweden, Index Copernicus International Ltd, Poland, Cite Factor, DOI: 10.15373/2249555XThomson Reuters"
- Aparna Goyal, Teena Bagga, Sanjeev Bansal, "MOBILE COMMERCE AN EMPIRICAL RESEARCH IN DELHI/NCR" INNOVATIVE JOURNAL OF BUSINESS AND MANAGEMENT 2277—4920 [2016] page © doi Thomson Reuters "under evaluation"
- Aparna Goyal, "Competition, Competitive Advantage and Role of Integrated Marketing Communications" International Journal of Computational Engineering Research 2250-3005 [2017] vol. 2 page329-336 © UGC approved, Google Scholar, DOAJ,
- Aparna Goyal, "Consumer Perception of the features and usage of Electronic Wallets" International Journal of Current Advanced Research ISSN: E: 2319-6475, ISSN: P: 2319-6505 [2017] Volume 07, Issue 08 page73-94 © UGC approved
- Dr. Aparna Goyal, "Categories of Influencing Factors for Consumer Purchase Behaviour of Innovative Environment-friendly Products: a study in Delhi/ NCR" Global Journal for Research Analysis 2277-8160 [2014] page © Google Scholar,

ULRICH'S PERIODICALS DIRECTORY, JOUR Informatice, ijIndex, INNO SPACE, Eye Source, ISI, World Cat, Index Copernicus, Open J Gate, DJOF, Cite Factor, GENAMICS, EZ3 Electronic Journal Library, ICI, Indian Science, Socol@r,DRjI,I2OR,COSMOS, CSA Illustration

- Aparna Goyal, "On line trends towards buying behavoiur of customers from different segments: Study in NCR:" International Journal of (IJPSR) ISSN (Online): 0975-8232, ISSN (Print): 2320-5148 [2017] page © Thomson Reuters, Web of Science - Emerging Sources Citation Index, PubMed (Selected citations), EMBASE (Elsevier), Corss Ref.,Hinari-WHO, Chemical Abstract, Scirus - Elsevier's, Gale- Expanded Academic ASAP, EBSCO, Google, Google scholar, International consortium for the advancement of academic publication (ICAAP), Scientific common, Pharmaceutical Sciences Open Access Resources (PSOAR), Index Copernicus, Ulrich's International Periodical Directory, ProQuest, New York University Health Sciences Libraries, Research Gate, Open-J-Gate, Geneva Foundation for Medical Education & Research, Ayush Research portal and Genamics Journal Seek. Efforts are ongoing to associate IJPSR with more of such renowned databases.

- Aparna Goyal, "New Product Launch Strategies and their Impact on Customers for Cosmetics Brand Organic Ayurveda India" World Journal of Engineering Research and Technology 2454-695X [2017] page © IISI, Cite Factor, BASE,CHS,IISS. Cold Spring harbor University.

- Dr. Aparna Goyal, Dr. Sanjeev Bansal, "Consumer Perception towards Natural Beauty Productsw.r.t. Environmental Strategies" International Journal of Engineering and Science Invention ISSN (Online): 2319 – 6734, ISSN (Print): 2319 – 6726 [2017] Volume 6 Issue 11 page46-56 © Google Scholor, EBSCO, Cabells Directories, Pro Quest, Open J Gate, Jour Informatics, SCINS, Ulrichs Web

Conferences – Presentations/ Proceedings:

- Aparna Goyal, "Eco Friendly Marketing: The market potential for sustainable products" Trinity Journal of Management, IT, & Media 2320-6470 [2014] Vol.5, issue 3 page53-57
- Dr. Aparna Goyal, Dr. Sanjeev Bansal, "Effectiveness of Digital Marketing in the Challenging Age: An Empirical Study" ABS- GLRC ISBN-978-93-86330-04-8 [2017] Vol.1, No.1Abstract page138
- Aparna Goyal, "Factors Affecting the Consumer Choice for Online Shopping over the In-Store Shopping" ABS- GLRC ISBN-978-93-86330-04-8 [2017] page108
- Aparna Goyal, "Study of factors leading to consumer motivation to switch towards an online purchase platform." Global Leadership Research Conference on Leadership Governance & Public Policy 2320-5407 [2016] page
- Aparna Goyal, "To study the consumer e-buying behaviour and strategies used by e-commerce websites, with reference to marketing of Apparels"

2nd Global Leadership Research Conference [2017] page65
- Aparna Goyal, "A study of factors affecting the decision of consumers in deciding the institution for housing loan" Global Leadership Research Conference on Leadership Governance & Public Policy [2016] page
- Aparna Goyal, "On Cutting Edge Research" International Conference on Innovation -In partnership with Imperial College Business School and Singapore Management University hosted by BML Munjal University 2278-7798 [2016] Vol6, issue 6 page160-163
- Dr. Aparna Goyal, Dr. Sanjeev Bansal, "Perceptual Mapping of Top 5 e-Retail Websites: With special reference to students in Delhi/NCR" Global Leadership Research Conference on Leadership Governance & Public Policy ISBN: 978-93-8500-48-5 [2015] page
- Aparna Goyal, "Social media marketing as platform for positive influence towards purchase of eco labelled green products" Global Leadership Research Conference on Leadership Governance & Public Policy [2016] ED-1 page330-356
- Aparna Goyal, "Strategic Human Resource Management Effectiveness As Determinants of Firm's Performance" Global Leadership Research Conference on Leadership Governance & Public Policy [2016] page64
- Aparna Goyal, ""Perceptual Mapping of top 5 online websites on various attributes of eco-friendly products: with special reference to Delhi/NCR", vol.1, ISBN:9789385000-18-5,pp.331-356, N"

National Seminar on Roots of Indian management [2015] page
- Aparna Goyal, "Ruby Tuesdays – Running Tonne Days, "Renvoi International case study collection" - ABS AUUP - [2005] Vol 5 page25-32
- Aparna Goyal, ""E- Commerce- The Biggest Revolution in Business" Details: Paper presented in XIV annual Conference of All India Association for Educational Research (AIAER) held in Ghaziabad and published in the" AIAER Conference Magazine 0970—9827 [2002] page
- Aparna Goyal, ""Study of Consumer Behavioral Eco-Patterns in Indian Retail Sector for Tractors in NCR", DU (Paper accepted presented, to be published in Jan 2016" Indian Retail Conference (IRC2016) during February 26-27, 2016 at IIC, Delhi organised by SBPPSE, AUD, Delhi [2016] http://ijpsr.com/articles/?iyear=91&imonth=72 pageICV
- Aparna Goyal, Ruby Tuesday - Running Tonne Days" Amity University [2005]
- Aparna Goyal, "Partridge Publishing" Partridge Publishing, 1663 Liberty Drive Bloomington, IN 47403[2017] page 150 "under review"
- Dr. Aparna Goyal, Dr. Sanjeev Bansal, "To Study the effect of employee training for new Franchise Employees at Body Shop in Influencing Environmentally Sustainable Strategies" International Conference on Family Business & Entrepreneurship [2017] page © SCOPUS
- Dr. Aparna Goyal, Dr. Sanjeev Bansal, "To study role of Intermediaries in Eco-Organic Food Market and its Effect on Consumer Perception of

Environmentally Green Products" International Conference on Family Business & Entrepreneurship [2017] page © SCOPUS
- Aparna Goyal, "A study on methods used for simplification of Mutual Funds Marketing Strategies in Indian scenario" Meerut Institute of Technology -International Conference on Advanced Research and Innovation in Engineering (ICARIE-2017) [2017] page © UGC approved
- Aparna Goyal, "Study of Content Marketing Strategies in Publishing Sector in Indian Context" Meerut Institute of Technology -International Conference on Advanced Research and Innovation in Engineering (ICARIE-2017) [2017] page © UGC approved
- Dr. Aparna Goyal, Dr. Sanjeev Bansal, "Eco-friendly Sensory Sustainable Environment CSR Marketing for Natural Organic Farming" Journal of Advanced Research in Dynamical and Control Systems 1943-023X [2017] Vol. 3, Issue 5 page363-378 © ELSEVIER, SCOPUS, UGC
- Dr. Aparna Goyal, "Understanding Environment Sustainability Quotient: Reasons & Challenges" Journal of Advanced Research in Dynamical and Control Systems 1943-023X [2017] page © ELSEVIER, SCOPUS, UGC
- Dr. Aparna Goyal, Dr. Sanjeev Bansal, "Study of Consumer Perception and Attitude towards Contemporary Technological Innovativeness w.r.t. Electronics Sector" Journal of Advanced Research in Dynamical and Control Systems 1943-023X [2017] page © ELSEVIER, SCOPUS, UGC

Under Publication:

- Dr. Aparna Goyal, "Three Pillars of Consumer Behaviour w.r.t Influence on Satisfaction and Delight" World Wide Journal of Multidisciplinary Research and Development ISSN: 2454-6615 (Online) [2017] page © Google Scholor, EBSCO, ISI, Index Copernicus International, Research Gate, e-journals.org, Scribd, Ulrichs Web,Pubshub, Cite Seer, UDL Theses, The Max Perutz Library
- Dr. Aparna Goyal, "Study of Internal and External Factors Affecting Level of Satisfaction in an Organization"" International Journal of Engineering Research and Development ISSN(Online): 2278-067X, ISSN(Print): 2278-800X [2017] page © UGC, Index Copernicus, Google Scholar, Informatics, ProQuest, Research Gate, Docstoc, Scribd, UlrichWeb, Internet Archive, Queen's University, Goethe University"
- Dr. Aparna Goyal, "Marketing Strategies w.r.t. Social Group Factors in Case of Innovative Products" American Scientific Research Journal for Engineering, Technology, and Sciences 2017 page Ulrich's, Massachusetts Institute of Technology (USA), Open Archives (Cornell University (USA)), Ulrich's Periodicals Directory, Simpson University (USA), IE Library (Spain), Tilburg University (The Netherlands), McGill University (Canada), INDIANA UNIVERSITY-PURDUE UNIVERSITY INDIANAPOLIS (USA), Indiana University East (campus library (USA)), University Of Arizona (USA), OCLC World Cat, University

Of Washington (USA), Biola University (USA), OAIster database.

- Dr. Aparna Goyal, Dr. Sanjeev Bansal, "A Study of the crucial issues faced in online Digital Marketing Platforms, with reference to the Viral and WOM Marketing Strategies." International Journal of Scientific and Research Publications 2320-5407 [2017] page ©
- Dr. Aparna Prashant Goyal, "Perceptual Influence on Banking Processes by using CSR through Ecologically Sustainable Innovative Practices" INTERNATIONAL JOURNAL OF ADVANCED RESEARCH (IJAR) ISSN: 2349-8862 [2017] page © Google Scholar, CNKI Scholar, DOAJ Sweden, Index Copernicus International Ltd, Poland, Cite FactorThomson Reuters
- Aparna Goyal, "Ecologically Sustainable Eco-Green Initiatives by Banks due to Awareness and Perception of Internal and External Customers with regard to Green" International Journal of Scientific Research and Engineering Studies 2349-8862 [2017] ` page © ICI JOURNALS MASTER LIST, Medical Council of India MCIThomson Reuters
- Dr. Aparna Prashant Goyal, "Sensory Marketing as an upcoming strategy in affecting Consumer Buying for Eco-Products" International Journal of Applied Engineering Research [2017]
- Dr. Aparna Goyal, "To understand the Buying Behaviour of Online Shoppers, w.r.t. Reliability and Safety" IJSER 2229-5518 [2017] 17-Vol.-5-Issue-12-December-2014-IJPSR-RA-4224-Paper-17 page ICV

- Aparna Goyal, "To Study the Impact of Branded Retail Store by way of its Transparency in Labelling and Packaging in Food Sector" American Journal of Engineering Research [2017]
- Aparna Goyal, "Study of Rise of Digital Marketing Strategies for Creating Environmental Awareness using CSR" International Journal of Engineering Sciences & Management 2231-3273 [2017] page © Google Scholar

www.ingramcontent.com/pod-product-compliance
Lightning Source LLC
Chambersburg PA
CBHW030758180526
45163CB00003B/1082